The Inventing of America

The Rawding family outside their sod house north of Sargent, Custer County, Nebraska, c. 1889

The Inventing of America

Bruce Norman

Taplinger Publishing Company
New York

First published in the
United States in 1976 by
TAPLINGER PUBLISHING CO., INC.
New York, New York

Library of Congress Catalog Card
Number: 76-11686

ISBN 0-8008-4220-0

Contents

Train of the Santa Fe Railroad crossing the Cañon Diablo, Arizona, c. 1903

Acknowledgements

For my Father and Mother
and for all
my American friends,
especially Freda and Seymour.

I would like to thank the Indiana Historical Society; the Conner Prairie Pioneer Settlement; the New York Public Library; the National Geographic Society; the American Life Foundation; the British Museum; Mrs Hedda Harrigan Logan for permission to quote from her father's unpublished sketch 'The Telephone'; Cromwell Music Ltd for permission to reproduce part of the song 'Old Man Atom'; Messrs Simon and Schuster for permission to quote from Mitchell Wilson's *American Science and Invention*; The Ronald Press Co., New York, for permission to quote from John W. Oliver's *History of American Technology*, copyright © 1956; Random House for permission to quote from D. J. Boorstin's *The Americans*; General Foods Corporation for permission to reproduce their Maxwell House commercial; and Faber and Faber for permission to reproduce the poem 'Death of a Toad'.

I would like to thank Professor Tom Hughes of the University of Pennsylvania for constructive suggestions; Charmian Campbell, Fanny Prior and Sally Evans for help with research; and my colleagues Lawrence Wade, James Burke and Michael Glynn; for editorial help with the production of the book, Diana Souhami, Barbara Fenton, and Peter Campbell. I would also like to thank the staff and students of Vallejo Senior High School (1964–5), and Mr and Mrs Darrel Catling for their hospitality.

Bruce Norman
January 1976

Introduction

The Safety Pin. The greatest little invention in the world. American. Formerly Ancient Greek. It was invented by Walter Hunt in 1849 to pay off a debt of $15. It took him three hours.

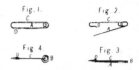

UNITED STATES PATENT OFFICE

DRESS PIN

Specification of letters Patent No. 6281 dated April 10, 1849

To all whom it may concern:

Be it known that I, Walter Hunt, of the city, county and State of New York, have invented a new and useful Improvement in the Make or Form of Dress Pins, of which the following is a faithful and accurate description. The distinguishing features of this invention consist in the construction of a pin made of one piece of wire or metal combining a spring, and clasp or catch, in which catch, the point of said pin is forced and by its own spring securely retained. They may be made of common pin wire or of the precious metals.[1]*

The description is accompanied by diagrams, and Hunt sold the patent rights for $400.

Hunt also invented a sewing machine, a repeating rifle, a revolver, a velocipede, a paper collar, a knife sharpener, a coal-burning stove, a machine for making nails, a machine for sweeping the street, paraffin candles, conical bullets, and a sea-camel (a sea-camel? A kind of dry-dock); an ice plow (British read plough), and a pair of shoes for a circus clown to play fly and walk up walls. He was one of the nation's most inventive inventors.

Hunt is just one of the stops on America's progress from Nowhere to the Moon in 200 years. From remote British colony to greatest industrial nation on earth. A sequence of invention and achievement unmatched in history. A series of startling firsts— firsts even though most of them did have their roots in Europe— the telegraph, the telephone, the screw-top jar, the lift (Americans read elevator), the sleeping car, barbed wire, the freezer, the gramophone, the Pill, the cocktail, the airplane and the Bomb. And dozens more.

American know-how has made the American Way of Life. Since the invention of the movies and, increasingly since the transistor made possible the satellite which made possible instant communication, it has made most other people's Way of Life as well. We may not like the Life. We may not want the invention. Neither the Life nor the inventions may be what the pioneer Americans intended. But now that we are stuck with them, how did they happen?

Walter Hunt, 1796–1859

*The footnote figures refer to the Book List on page 232.

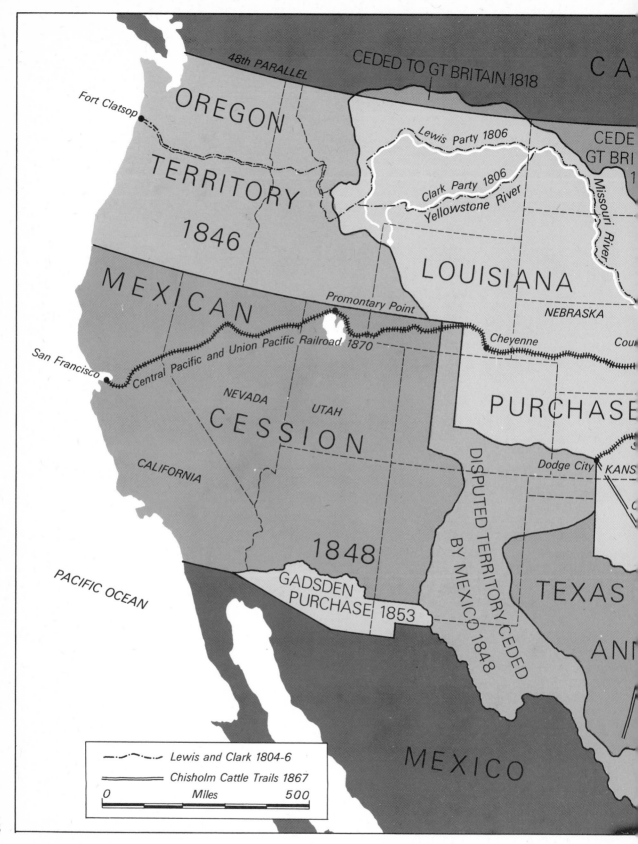

48th PARALLEL

CEDED TO GT BRITAIN 1818

CA

OREGON

Fort Clatsop

CEDE
GT BRI
1

Lewis Party 1806

Clark Party 1806

TERRITORY

Yellowstone River

Missouri River

1846

LOUISIANA

MEXICAN

Promontary Point

NEBRASKA

San Francisco

Central Pacific and Union Pacific Railroad 1870

Cheyenne

Cou

PURCHASE

NEVADA

UTAH

CESSION

CALIFORNIA

Dodge City

KANS

1848

DISPUTED TERRITORY CEDED BY MEXICO 1848

TEXAS

GADSDEN
PURCHASE 1853

PACIFIC OCEAN

AN

MEXICO

Lewis and Clark 1804-6

Chisholm Cattle Trails 1867

0   Miles   500

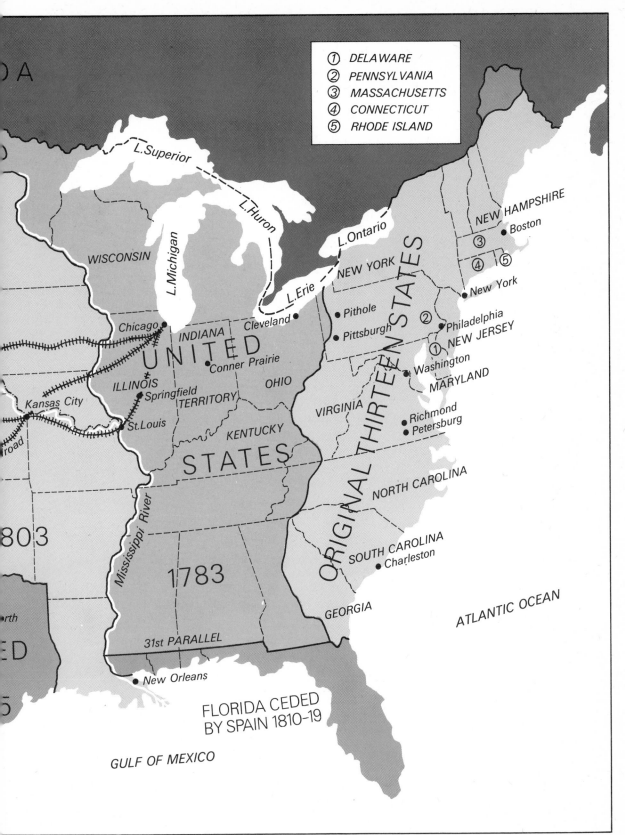

DELAWARE

① DELAWARE
② PENNSYLVANIA
③ MASSACHUSETTS
④ CONNECTICUT
⑤ RHODE ISLAND

L.Superior

L.Huron

L.Michigan

L.Ontario

NEW HAMPSHIRE

② Boston

WISCONSIN

NEW YORK

L.Erie

③

④ ⑤

New York

Chicago

Pithole

②

INDIANA

Cleveland

Philadelphia

UNITED

Pittsburgh

① NEW JERSEY

Conner Prairie

Washington

ILLINOIS

OHIO

MARYLAND

Kansas City

Springfield

VIRGINIA

Richmond

St.Louis

TERRITORY

Petersburg

KENTUCKY

STATES

NORTH CAROLINA

road

Mississippi River

1803

SOUTH CAROLINA

1783

Charleston

GEORGIA

ATLANTIC OCEAN

31st PARALLEL

New Orleans

FLORIDA CEDED
BY SPAIN 1810-19

GULF OF MEXICO

ORIGINAL THIRTEEN STATES

9

Chapter 1

'The Union Manufactories of Maryland'—a pen and ink drawing by Maximilian Godefroy

The State of the Nation

On Patapsco-Falls Baltimore County

The thirteen 'United States of America', who declared their independence from Britain on 2 July 1776, were united against only two things – the rule of King George III and the 'forest'.

King George and his Redcoat emissaries were a tangible enemy, the 'forest', or 'wilderness', was not. Awe-inspiring, frightening, unknown, it stretched from the inland boundary of the new nation's toe-hold on the Atlantic seaboard westward as far as anybody cared to speculate. Forests, rivers, giant mountains—the haunts of wild animals and wild Indians.

It had been 150 years since the first Europeans settled in Virginia, and the new nation was now strung out along the Atlantic coast for a thousand miles. They were no more than two and a half million people, fewer than lived in London. There were only a few large towns: Charleston (10,000), Boston (17,000), New York (20,000) and Philadelphia (25,000), all on the coast. Communications were appalling: there were few roads and no canals. Transport was easiest by sea.

From a distance, the towns looked like miniature versions of eighteenth-century London, with tall church steeples and pink or white brick houses. On closer inspection, the houses proved to be mostly of wood and roughly constructed—urban versions of the settlers' log cabins. The overwhelming impression was of space: 'the staple commodity is Land . . . the appearance everywhere of a vast outline with much to fill up'.

Because these communities were so widely scattered, they each retained 11 many of the traditions of their countries of origin. And, in consequence, those countries' mutual dislike and suspicion. The Dutch disliked the English, the English disliked the Germans, and everyone disliked the Virginians who thought themselves superior because they'd got to America first.

Benjamin Franklin, only six years before the rebellion, saw little danger of a united opposition to the mother country:

If they could not agree for their defence against the French and the Indians who were perpetually harrassing their settlements, burning their buildings and murdering their people: can it be reasonably supposed there is any danger of their uniting against their own nation, which protects and encourages them, with which they have so many connections and ties of blood, interest and affection and which 'tis well known they love much more than they love one another?[6]

The martial marriage of these DisUnited States was purely one of convenience. After the ceremony was successfully concluded and the Founding Fathers had proclaimed the importance and dignity of each human being (particularly male), the partners squabbled. There were fisticuffs in Congress and the States began to drift apart. What finally held them together and made them into a nation was the surviving common enemy—the Wilderness—and their common background, Europe. In the early stages it was Europe that provided the fundamental inventions which, suitably adapted and applied, were to enable the Americans to overcome that hostile and alien environment.

These early Americans were independent, practical, hardworking, puritanical, simplistic in outlook and anti-intellectual—which does not necessarily mean they were anti-education. In Massachusetts, fathers were fined for not sending their children to school. In a largely classless society, free from the poverty that characterised much of Europe, the colonist earned his living from farming and, especially in New England, from the sea—fishing and trading. Other industries, where they existed at all, were

Mid-18th-century view of the City of Boston. The proximity of the Wilderness is shown by the Indian family on the left and the deer crouched under the tree.

Below: a south prospect of 'Ye Flourishing City' of New York 1746. 18th-century views of American cities emphasised the harbour and the substantial buildings in the hope of attracting settlers and capital.

primitive—printing, tanning, paper-making. Most manufacturing was done at home and there was little use of machinery.

Colonies, in Britain's eyes, existed for two things: to supply raw materials and, in return, be supplied with manufactured goods. Self-improvement was discouraged except where it was to Britain's benefit. America constructed over one-third of Britain's shipping and had a thriving overseas trade, but she was not permitted to manufacture nails—for sale. So, determined not to buy British, the colony was criss-crossed by farmers' fences made out of plaited branches.

As a result of what Adam Smith, in the year of Independence, called Britain's violation 'of the most sacred rights of mankind'[1]— the right of the colonists to do their own thing—America was singularly unprepared for national, economic survival. There was no native tradition of technical know-how, nor of scientific research on which they could build. With one exception.

Benjamin Franklin (1706–90), like Shakespeare, was 'not for an Age but for all Time'. The grandfather of American invention, he was a renaissance man, equally at home in politics, the arts or the sciences. A man with a mind able to penetrate to the basic simplicity behind apparent confusion and complexity. William Watson, a fellow member of the British Royal Society, called him 'a very able and ingenious man who had a head to conceive and a hand to carry into execution'.[2] A man who, in his declining years, divided his time between structuring the Constitution of the United States and inventing bifocals in order to see to do it properly.

Finding this change [between pairs of glasses] troublesome, and not always sufficiently ready, I had the glasses cut, and a half of each kind associated in the same circle. By this means, as I wear my spectacles constantly, I have only to move my eyes up or down, as I want to see far or near, the proper glasses being always ready. [1784][6]

He was in the great tradition of European Natural Philosophy— the eighteenth-century term for 'science'—and his reputation in Europe was that of the Isaac Newton of electricity. Today's terms 'minus', 'plus', 'positive', 'negative' and 'armature' were his, and he formulated a workable theory of electricity's nature, namely that it was composed of 'subtle particles' that could penetrate substances like metal and glass and could be redistributed. It was well over another hundred years, not until 1897, before the electron was discovered.

In 1743, Franklin was in Boston.

I met there with a Dr Spence, who was lately arrived from Scotland

Benjamin Franklin, 1706–1790

and showed me some electric experiments. They were imperfectly performed, as he was not very expert: but, being on a subject quite new to me, they equally surprised and pleased me.[7]

Back in Philadelphia:

I eagerly seized the opportunity of repeating what I had seen at Boston and by much Practice acquired great Readiness in performing these ... adding a Number of New Ones. I say much Practice, for my House was constantly full for some time, with People who came to see these new Wonders.[7]

Franklin worked especially with a new device called the Leyden jar, just a corked bottle of water which, when charged, mysteriously caused an enormous build-up of electricity. Franklin's question was simple: what caused the build-up? Joseph Priestley, in his 'History and Present State of Electricity' (1767), explained the Franklin experiments to the British:

In order to find out where the strength of the charged bottle lay, he placed it upon glass; then first took out the cork and the wire, and, finding the virtue was not in them, he touched the outside coating with one hand, and put the finger of the other into the mouth of the bottle: when the shock was felt quite as strong as if the cork and wire had been in it. He then charged the phial again, and pouring out the water into an empty bottle insulated, expected that if the force resided in the water it would give the shock; but he found it gave none. He then judged that the electric fire must either have been lost in the decanting, or must remain in the bottle; and the latter he found to be true; for, filling the charged bottle with fresh water, he found the shock and was satisfied that the power of giving it resided in the glass itself.[10]

Franklin did not stop there.

Upon this principle, Dr Franklin constructed an electrical battery consisting of eleven panes of large sash glass, coated on each side, and so connected, that charging one of them would charge all. Then, having a contrivance to bring the giving sides into contact with one wire, and all the receiving sides with another, he united the force of all the plates, and discharged them all at once.[10]

It was the world's first condenser. But:

The greatest discovery which Dr Franklin made concerning electricity, and which has been of the greatest practical use to mankind, was that of the perfect similarity between electricity and lightning. . . .[10]

1 Flashes of lightning, he begins with observing, are generally seen crooked and waving in the air. The same, says he, is the electric spark always, when it is drawn from an irregular body at some distance.

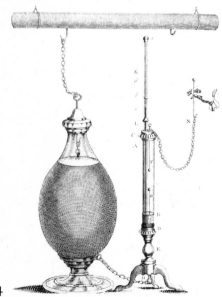

The Leyden jar. The spark was produced across gap F-G and the subsequent rise in temperature measured by the thermometer.

2 Lightning strikes the highest and most pointed objects in its way preferably to others, as high hills and trees, towers, spires, masts of ships, points of spears etc. In like manner, all pointed conductors receive or throw off the electric fluid more readily than those which are terminated by flat surfaces.

3 Lightning is observed to take the readiest and best conductor. So does electricity.

4 Lightning burns. So does electricity.

5 Lightning sometimes dissolves metals. So does electricity.

6 Lightning rends some bodies. So does electricity.

7 Lightning has often been known to strike people blind. And a pigeon, after a violent shock of electricity, by which the doctor intended to have killed it, was observed to be struck blind likewise. . . .[10]

Franklin recorded his observations in 1749. The following year, he suggested an experiment that might prove them.

On top of some high tower or steeple place a kind of sentry box . . . big enough to contain a man and an electrical stand [an insulated platform]. From the middle of the stand let an iron rod rise and pass bending out of the door, and then upright twenty or thirty feet, pointed very sharp at the end. If the electrical stand be kept clean and dry, a man standing on it when such clouds are passing low might be electrified and afford sparks, the rod drawing fire to him from a cloud.[6]

What were known as the 'Philadelphia Experiments' were actually performed successfully, to much amazed reaction, and—equally amazing—with no fatality, by a French scientist, D'Alibard, near Paris, in May 1752. Franklin, across the Atlantic, unaware of what had happened and himself still waiting for someone to build a suitably high tower or steeple, thought of another experiment.

To demonstrate, in the completest manner possible, the sameness of the electric fluid with the matter of lightning, Dr Franklin, astonishing as it must have appeared, contrived actually to bring lightning from the heavens, by means of an electrical kite, which he raised when a storm of thunder was perceived to be coming on.[10]

15

The kite experiment. The first great American scientific legend.

Preparing, therefore, a large silk handkerchief and two cross-sticks of a proper length on which to extend it, he took the opportunity of the first approaching thunderstorm to take a walk in the fields, in which there was a shed convenient for his purpose. But, dreading the ridicule which so commonly attends unsuccessful attempts at science, he communicated his intended experiment to nobody but his son who assisted him in raising the kite.

The kite raised, a considerable time elapsed before there was any appearance of its being electrified. One very promising cloud had passed over it without any effect; when, at length, just as he was beginning to despair of his contrivance, he observed some loose threads of the hempen string to stand erect, and to avoid one another, just as if they had been suspended on a common conductor. Struck with this promising appearance, he immediately presented his knuckle to the key, and (let the reader judge of the exquisite pleasure he must have felt at that moment) the discovery was complete. He perceived a very evident electric spark. Others succeeded, even before the string was wet, so as to put the matter past all dispute, and when the rain had wet the string he collected fire very copiously. This happened in June 1752.[10]

That the English Joseph Priestley, one of the world's first and foremost chemists, should have commented at such length on Franklin's achievements is mark of the American's great standing. Priestley called the discovery 'the greatest, perhaps, that has been made in the whole compass of philosophy since Newton'. No mean achievement for a colonialist cut off by a thousand miles of ocean from the main centres of scientific thought. But this isolation proved an advantage. Franklin was forced to rely on his own observations and not on European hypothesis. When, after the 1750s, he was regularly in touch with European thinking, his fundamental discoveries in electricity came to an abrupt end.

Franklin, the 'pure' scientist, was as much concerned with science 'applied'. This philanthropic regard for the welfare of one's fellow human beings, which devolved from the Royal Society, was to be the concern of American science for the next 150 years. Referring to his electrical experiments, Franklin says he was 'chagrined' that he had not yet been able to produce anything 'of use to mankind'. Not quite true, as he frequently electrocuted turkeys for his dinner—but this was, perhaps, not quite what he had in mind. In 1753 he observed:

It has pleased God in His goodness to mankind at length to discover to them the means of securing their habitations and other buildings from mischief by thunder and lightning.[8]

He was referring to his invention of the lightning rod (British read conductor)—the direct spin-off from his kite experiments. By 1755 conductors were already in use in America, though not unanimously approved of:

God casts his thunderbolts where He lists, and it is presumption in man to endeavor to turn them aside.[11]

Franklin was pleased to see conductors in use in London when he visited in 1772, despite similar religious opposition there.

Equally the result of science 'applied' was the Franklin Stove of 1740. It developed out of his study of the physics of heat. He noted that air, heated, becomes rarefied, and rises. When it does so, a fire produces its greatest heat. Result—the 'Pennsylvania Fireplace' (the Franklin Stove). Instead of letting that heat escape up the chimney, the fireplace trapped warm air in an air box at the rear and allowed it to escape into the room from shutters at the side. But Franklin refused to patent his invention:

I declined it from a principle which has ever weighed with me on such occasions [that] as we enjoy great advantages from the inventions of others, we should be glad of an opportunity to serve others by an invention of ours, and this we should do freely and generously.[7]

It was a gesture rarely to be repeated in the subsequent history of American invention. For the 'sake of mankind', Franklin agreed to the stove's manufacture. His marketing commercial set a brilliant precedent for future businessmen. He listed the fireplace's considerable benefits:

People need not crowd so close round the Fire but may sit . . . with comfort in any Part of the Room, which is considerable Advantage in a large Family, where there must often be two Fires kept, because all cannot conveniently come at one.

He also dismissed the rival 'improved' fireplaces:

. . . [the draught] rushed in at every crevice so strongly as to make continual whistling and howling: and 'tis very uncomfortable as well as dangerous to sit against any such crevice.

He appealed to female vanity:

Women, particularly, from this cause (as they sit much in the house) get colds in the head, rheums and defluxions, which fall into their jaws and gums, and have destroyed early as many a fine set of teeth in these northern colonies.

And delivered the knock-out blow:

Great and bright fires do also much contribute to damage the eyes, dry

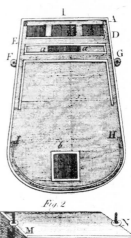

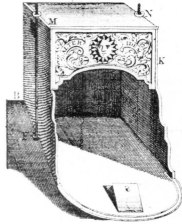

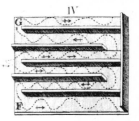

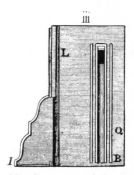

The Franklin Stove—people 'may sit with comfort in any Part of the Room'

and shrivel the skin, and bring on early the appearance of old age.[9]

In short, if you don't buy the 'Pennsylvania' you'll grow ugly, senile, and probably die.

Just as his experiments in Electricity were to point forward to Henry, Maxwell, Edison and Langmuir—to the telephone, the electric light bulb, radio, television and electronics—so his stove was the forerunner of central heating. Single-handed, Franklin set the pattern for scientific and technological progress in the United States. But advances in 'pure' science were not to come again, in any significant proportion, for almost a century. The immediate future for America lay neither with 'pure' science nor, even, with science 'applied', but with technology: with what the Penguin English Dictionary terms 'the science of the industrial and mechanical arts', to which Webster adds 'industrial science—especially of the more important manufactures as spinning, weaving, metallurgy, etc.' In other words, with machines.

Her independence won, America had suddenly to shift for herself. After a short period of euphoria, it became obvious that the 'opportunity' in the Land of Opportunity was the opportunity for men to solve the new nation's problems. And that required men like Eli Whitney (1765–1825), 'who', recorded a contemporary, 'out of respect for his distinguished talents, his private virtues and his public spirit, holds one of the highest places amongst the benefactors of our country.'

Whitney began young. According to his sister Elizabeth:

Eli Whitney (1765–1825), painted by the inventor of the telegraph, Samuel Morse

In the time of the revolutionary war, nails commanding a high price, Eli proposed having a forge . . . and making nails. Father consented. The forge was put up and Eli made nails, did his day's work and gained time to make tools for his own use . . . it was the second winter after he had begun to make nails he told me he laid a plan to hire a man, pay his board and have the profit of his work.[14]

Nails were still in short supply following the British interdiction; Whitney was just in his teens. His ability to provide what the nation wanted at a time when it most wanted it was to characterise Whitney's career. And the drive, the initiative, the ingenuity and the ultimate financial success made him into the archetype of the successful American.

Whitney's first sphere of operation was the South. Already a land of large plantations, colonial mansions and slaves, the South's wealth depended on rice, tobacco, indigo and – recently introduced – cotton. Rice was ruining the soil, the tobacco market was exhausted, and slavery was in decline. With cotton as the only potential money spinner, and with the cotton industry in the

North scarcely begun, the plantation owners had either to increase the amount of cotton for sale to England or go bankrupt.

Most of the cotton was of the sea-island variety. Introduced in 1786 from the Bahamas, its great advantage was that its black seed could be easily separated from the long hair. Its disadvantage was that it could not and did not grow anywhere else but on the forty-mile strip of the Georgia-Carolina coast. The alternative plant was green-seed cotton, which grew in the uplands. The problem with that was that it took ten hours of hard labour to produce a mere pound of seed-free cotton.

In 1793, at the age of 28, Eli Whitney accepted a job as a private tutor in South Carolina. But, on arrival, disappointed in the money he expected, he quit. At the invitation of the widow of General Nathanael Greene, he stayed on her plantation to help her manager.

Whatever he undertook . . . [he] seemed to have the Sagacity to perceive the possible consequences.[14]

Seeing cotton for the first time, Whitney perceived the consequences very quickly. He wrote to his father on 11 September:

I have . . . heard much said of the extreme difficulty of ginning cotton. That is, separating it from its seed. There were a number of respectable Gentlemen at Mrs Greene's who all agreed that if a machine could be invented which would clean the cotton with expedition, it would be a great thing both to the country and the inventor. . . . I involuntarily happened to be thinking on the subject and struck out a plan of a machine in my mind.[14]

Cotton plantation in the early 19th century. The owner's mansion is on the right, slave quarters are on the left. The white overseer carries a whip.

Whitney did not invent the cotton gin. There already existed a gin for de-seeding sea-island cotton. It looked like a mangle (Americans read clothes-wringer) and had two ribbed rollers. The cotton was fed in one side and wound out the other, the friction of the rollers being sufficient to squeeze the seeds out. But friction alone was not enough to squeeze out green seeds. These had to be laboriously picked out by hand. Whitney's innovation was to design a machine that duplicated these hand movements.

I made a little model for which I was offered, if I would give up all right and title to it, 100 guineas.[13]

Metal slats, acting like a wire mesh, held the cotton back from the rest of the machine, but allowed the cotton fibres to poke through. A drum, covered with fine, finger-shaped hooks, was rotated past the mesh and clawed out the cotton fibres. The seeds fell away as they knocked against the mesh, whilst the separated cotton lint on the drum was cleaned off by a rotating brush. The design had the simplicity of genius.

This machine may be turned by water or horse, with the greatest ease and one man will do more than fifty men with the old machines. It makes the labour fifty times less without throwing any class of people out of business.[14]

One class of people not put out of business was the slaves. Slavery revived along with the cotton industry. Before Whitney had even patented his machine, the farmers were planting green-seed cotton; production in South Carolina multiplied twelve times in ten years. By 1795 six million pounds were being exported yearly to the mills of Manchester. British capital poured into the South. Lord Macaulay pronounced:

What Peter the Great did to make Russia dominant, Eli Whitney's invention of the cotton gin has more than equalled in relation to the power and progress of the United States.[2]

But there can be no technological change without a complimentary social one. It's a lesson the nation was slow to learn, perhaps never learnt, and Macaulay did not live to see how the gin, in solving a technical problem, only created a bigger one for society.

As cotton spread west, ruining the soil as it went, the South, by 1850, came into direct confrontation with the expanding North. The issues were the moral one of slavery and the economic one of poor husbandry. In the long run, the gin's effect on the South was to inhibit progress. Without cotton, the South could have found other ways of diversifying its economy. As it was, it became locked into a system, technological and social, from which it would only be released by the violence of Civil War.

Eli Whitney's cotton gin of 1793

Eli Whitney's cotton gin in use

Invention, as the gin illustrates, is rarely the product of a single mind working in isolation. Rather it is the product of one or several minds developing an idea that is already in embryo. The gin problem was there for whoever could solve it and, although Whitney is given the accolade 'Father of Invention', other men produced gins at the same time. One of them, Hodgen Holmes, invented a similar machine but with circular, serrated-edge saws in place of the wire fingers. Whitney, too, originally thought of using saws. Perhaps he came to the idea independently, or perhaps he knew Holmes's invention and just got his patent in first—gentlemanly conduct is not an attribute of the invention game.

Whitney was one of the first who saw invention as a way of making money—an American characteristic, if not peculiar to Americans. It was encouraged by a nation that stressed success, self-help and non-interference from the government (although the government did, on occasions, give financial support). But in those early days there was also a conflicting notion that inventions should be common property for shared, not individual good. The gin was pirated. The new Patent Act of 1793 stressed that proof of invention and proof of infringement was the burden of the inventor. Such proof was difficult and costly to obtain.

The use of this machine being immensely profitable to almost every individual in the Country, all were interested in trespassing and each justified and kept the other in countenance.[13]

So Whitney wrote to Fulton in 1811, after the gin had become public property. At one time, he had over sixty lawsuits pending. He made a piddling $90,000, scarcely enough to cover his legal costs, while the planters earned ten million dollars a year.

I have labored hard against the strong current of Disappointment which has been threatening to carry us down the Cataract of Destruction—but I have labored with a shattered oar and struggled in vain.[13]

In 1795 Whitney's gin factory in New Haven, Connecticut, burned down. Heavily in debt, but refusing to give way to despair, he left his partner to fight the legal battles in the South and focused his attention on the North. New England, the cradle of America's Industrial Revolution, was itself still in the cradle. Over the next half century, however, it was to produce all the ideas that were to shape America into an industrial nation.

The problems facing the North, and particularly New England, were different from those of the South, but stemmed from the same cause. America had Independence without being independent. As the South relied for prosperity on British capital, so the North, with little industry and less know-how, was forced to rely for manufactured goods on imports from Europe. The dangers of this dependency were brought home with a bang—literally, during the undeclared war against Napoleon's France in 1798. How do you fight a war, let alone win it, when your guns are made by the enemy?

In May 1798, Eli Whitney, from his rebuilt but unproductive Connecticut factory, again exercised his 'Sagacity to perceive the Consequences', both for the nation and himself. He wrote to the Secretary to the Treasury, Oliver Woolcott:

I have a number of workmen and apprentices whom I have instructed in working wood and metals and whom I wish to keep employed. . . . These circumstances induced me to address you and ask the privilege

Eli Whitney's Gun Factory, New Haven,
Connecticut, c. *1820*

of having an opportunity of contracting for the supply of some of the articles which the US may want. I should like to undertake the Manufacture of ten or fifteen thousand stand of arms.[14]

Whitney had never made a gun in his life—but he had a novel idea of how he might do it.

I am persuaded that Machinery moved by water adapted to this Business would greatly diminish the labor and facilitate the Manufacture of this article. Machines for forging, rolling, boring, grinding, polishing etc. may all be used to advantage.[14]

Whitney's belief, probably gleaned from practical experience of making gins, was that, given the right equipment and some kind of system, you could make anything. His emphasis on machinery is significant. His major problem, which was also one of America's, was a shortage of skilled manpower—Britain forbade the emigration of all skilled artisans. Whitney's system was an attempt to overcome that shortage.

In Europe a gun, whether made entirely by hand or with the aid of a machine, was fashioned by one skilled craftsman. The Whitney system replaced the skilled craftsman. By breaking the gun down into several parts, it became possible to make each part separately and in large numbers, and by making each component part 'as much like each other as the successive impressions of a copper plate engraving',[4] the parts were rendered interchangeable (any one part could fit any gun) and assembly made easy. No need, when the gun broke, to order a new one—just a new part. But, most important, each manufacturing process could be undertaken by men who were not, in the European sense, skilled enough to make a complete gun. Whitney's primary aim was,

. . . to substitute correct and effective operations of machinery for that skill of an artist which is acquired only by long practice and experience.[4]

What Whitney had devised were the component parts later to be assembled into the vehicle of mass-production. It was to open a whole new social as well as technological prospect. Fast turn-out, machine-made goods, cheap. Without mass-production the American standard of living, and that of the entire Western World, would not be possible. Whitney, of course, was oblivious to the long-term implications. He was merely a manufacturer with a particular problem, searching for the most profitable solution. But while the theory was relatively simple to expound, putting it into practice was not.

By 1801, Whitney had honoured none of his $134,000 contract.

He appeared before a sceptical Congress to try and convince them that he was not merely a plausible cheat. He presented them with a wide selection of the mechanical parts of a musket and invited them to assemble the parts into a working gun. Thomas Jefferson was convinced:

He has invented moulds and machines for making all the pieces of his lock so exactly equal that take 100 locks to pieces and mingle their parts and the 100 locks may be put together as well by taking the first pieces that come to hand.[14]

But the sample musket locks that Whitney displayed were still hand-made. If interchangeability was to work, even on a small scale, machine tools of considerable sophistication were essential: 'Tools [that] themselves shall fashion the work.'[16] Britain already had them, but they weren't for sale. The Colonies wanted independence—now they'd got it. 'No design, pattern, model or specification on any machine whatsoever' was to leave Britain. Penalty, £500 or twelve months in gaol. America, as Whitney realised, must make her own machine tools.

The first all-American milling machine was reputedly made by a man called Johnson in Middletown in 1808. It possessed a system of guides, patterns, template gauges and jigs specifically designed for use by unskilled workers. By 1812 Whitney himself claimed to have a machine that was able to 'guide' a man, rather than the man the machine. By 1818 Thomas Blanchard invented a 'copying' lathe which was designed so that it could reproduce the irregular shapes of a gun stock. By 1820 the British reported:

The Americans seem eager to resort to machinery wherever possible.

And the Americans crowed:

In twenty years, we've designed more machines than the whole of Europe put together.

Just as he was not the only inventor of the gin, Whitney was not the only American gun manufacturer operating the Whitney System. He might not even have been the first. Simeon North had an arms factory at Middletown Connecticut, only twenty miles from Whitney's. In 1799 he had a government arms contract that he completed in a year. By 1813 he had made 10,000 guns in ten years, and Whitney mentions luring men away from North and 'borrowing' their ideas. Whitney also visited John Hall's factory to see 'the new system being adopted there', and Hall's government contract specifically mentions 'interchangeability'. But this was 1820 and Hall was probably borrowing from Whitney, not the other way round. Perhaps, in a more idealistic age, these borrowings were less industrial espionage and more the free exchange of technical know-how in the national interest.

Whitney did not complete his guns contract until ten years after it was originally signed. His profit was a mere $2500. The big profit was to be America's. Ironically, the Whitney system, later the Uniformity system, and finally, to Europeans, the American System, wasn't American at all, but European.

In Paris Le Blanc had invented a gun with interchangeable parts as early as 1788. Jefferson, after the famous 1801 meeting in Congress, commented:

Le Blanc has extended his process to the barrel, mounting and stock. Mr Whitney has not yet extended his improvement beyond the lock.[13]

Samuel Bentham, in the Royal Navy docks at Portsmouth, had invented a system of forty-four separate machines to make pully blocks. It enabled unskilled men—prisoners—to do the work of 130 craftsmen. That the idea took root in America and not in Europe illustrates many of the differences that existed between them, differences which were eventually to end in Britain's industrial decline.

The British were traditionalist, craftsman-orientated, and resistant to change. With their superior machine tools they continued to turn out craftsmen-made objects for a social élite. The Americans, without tradition, seized on any new idea that could be useful. Their innovation was not in the design of their tools but in the way they used them to turn out an increasing number of uniformly made objects available for everyone. That the object was inferior to its British counterpart was irrelevant. If it worked, it was good enough.

It was no accident that the American Industrial Revolution began in New England. Its seafaring and trading traditions made the people adventurous and willing to experiment. The area also provided an abundance of natural resources—wood and water power and a growing market in the small towns and villages. It was here that one of the essential steps in the development of mass-production, one which Whitney himself took over—the concept of the factory—was first realised.

Samuel Slater (1768–1835), with the inflated title 'Father of American Manufactures', was not American. As a young boy in Derbyshire, England, he was apprenticed to a partner of the famous inventor Richard Arkwright. At the age of 21, against the law and without telling his mother, he slipped out of the country. In his head were the latest designs of Britain's cotton machinery: Hargreave's spinning jenny, Crompton's spinning mule and Ark-

Samuel Slater, 1768–1835

25

The Slater Mills, Pawtucket, Rhode Island, c. 1815. Like all New England factories it was built beside running water. Slater later said of his factory, 'I suppose that I gave out the psalm and they have been singing the tune ever since'.

wright's water frame. He reproduced them from memory and, in December 1790, three years before Whitney went South, Slater began to produce cloth in what was to become the first successful factory in America.

For several years the Slater Mill at Pawtucket, Rhode Island, remained an isolated phenomenon. There had been other mills, but they had failed. Slater set American industry an early example of the importance of business management. But in the context of mass-production he did much more than that. The mill was almost a carbon copy of its English progenitor; it broke down its manufacturing processes into distinct stages, each a simple task that could be done by a child. And if by children, then why not unskilled adults? And if for cotton, then why not the gun? Slater's breakdown of manufacture into a division of labour (1790) was the first step towards Eli Whitney's idea of interchangeability (1800). Without division of labour, interchangeability was unworkable.

Throughout the early years of the nineteenth century, the cot-

Carding, drawing, roving and spinning, as introduced by Samuel Slater in the Old Slater Mill, 1790

The mill city of Lowell, Massachusetts, on the Merrimack River, c. 1830

ton industry and the arms industry continued to develop together —each borrowing ideas from the other. Fresh impetus for both industries came in 1811 with Francis Lowell (1775–1817). Financier, administrator, speculator, he saw there was money to be made in machines. In Manchester on a rest cure he copied the designs of the new machinery (or as his partner, Nathan Appleton, phrased it, 'obtained all the information that was practicable') and smuggled them back to America. It was twenty years since Slater had done a similar thing.

Stealing British trade secrets had become something of an American patriotic pastime. And an unpatriotic British one. Slater and Lowell were not alone. George and Isaac Hodgson, Manchester machinists, were lured to the US by American blandishments in 1811. To avoid detection and prosecution, they sent their tools on a separate boat in cases labelled 'Fruit Trees'.

Lowell, himself safely arrived back in America, had his machines redesigned to make them even more labour-saving. He founded the Boston Manufacturing Company, the first factory in America to have all its machinery driven from a central power source—water—and 'where the power loom was first brought into successful operation on this side the Atlantic'. The design of the British power loom had been hawked round America in 1814 by a patriot (or traitor, according to your nationality) called Gilmore. Slater's factory did no weaving, which was 'put out' as it was in England, and Slater turned the offer down. Lowell snatched it up.

[Mr Lowell] invited me to go out with him and see the loom operate. . . . I well recollect the state of admiration and satisfaction with which we sat by the hour watching the beautiful movement of this new, wonderful machine destined, as it evidently was, to change the character of all textile industry.[19]

It was a breakthrough. The vision of a total cloth-making process, including weaving, and carried out entirely under one roof, could now become a reality. A production line, with green-seed cotton from the South in at one end and finished cloth out at the other. It was the birth of modern factory organisation; and it was to change America from a rural to an urban society.

In 1789 it had seemed that the future economic well-being of America lay with agriculture. Four years earlier Jefferson had written:

Cultivators of the earth are the most valuable citizens: As long therefore, as they can find employment in this line, I would not convert them into . . . artisans, or anything else. . . . I consider the class of artificers as the panders of vice and the instruments of which the liberties of a country are generally overturned.[22]

After the 1812 war with Britain he had changed his mind:

To be independent for the comforts of life, we must fabricate ourselves. . . . He, therefore, who is now against domestic manufactures must be for reducing us, either to dependence on that nation [England] or to be clothed in skins, and live like wild beasts in dens and caverns. I am proud to say that I am not one of these. Experience has taught me that manufacturers are now as necessary to our independence as to our comfort.

But there was a certain puritanical concern about factories—even from the owners themselves. William Gregg in 1811:

The moral standards of our community must not be impaired. Cotton factories should not be located in cities. There it would be impossible to control the moral habits of the operatives—and to keep up a steady, efficient and cheap working force.

Though the concern seemed as much for maintaining profits as for maintaining morals. But there were other, related worries:

The introduction of the cotton manufacture in this country, on a large

Mule spinning, c. 1835

scale, was a new idea. What would be its effect on the character of our population was a matter of deep interest. The operatives in the manufacturing cities of Europe were notoriously of the lowest character for intelligence and morals. The question therefore arose, and was deeply considered, whether this degradation was the result of the peculiar occupation or of other and distinct causes.[19]

America was in danger of reproducing the dark, satanic mills of England along with its dark, satanic workers. Lowell, borrowing most of his ideas from Robert Owen and the New Lanark Experiment, made a deliberate attempt to avoid the Dark Ages of European Industrialisation.

Here was in New England a fund of labor, well-educated, virtuous. It was not perceived how profitable employment has any tendency to deteriorate the character. The most efficient [safe-]guards were adopted in establishing boarding houses, at the cost of the Company, under the charge of respectable women, with every provision for religious worship. Under these circumstances, the daughters of respectable farmers were readily induced to come into the mills for a temporary period.[19]

Lowell, for all the high-minded talk, was making a virtue out of necessity. In an area where men preferred to work on their own land, labour was restricted to women and children anyway. But, by deliberately attracting, if only for a short period, a 'constant supply of female hands from the country', he could avoid creating the 'degraded' class of employee that was found in England. By careful if rather rigid supervision, he could enhance the dignity and value of the employee for the benefit of the whole community.

REGULATIONS TO BE OBSERVED BY PERSONS OCCUPYING THE
BOARDING HOUSES BELONGING TO THE
MERRIMACK MANUFACTURING COMPANY*

They must not board any persons not employed by the company, unless by special permission.

No disorderly or improper conduct must be allowed in the houses.

The doors must be closed at 10 o'clock in the evening; and no person admitted after that time, unless a sufficient excuse can be given.

Those who keep the houses, when required, must give an account of the number, names and employment of their boarders; also with regard to their general conduct, and whether they are in the habit of attending public worship.

The buildings, both inside and out, and the yards about them, must be kept clean, and in good order. If the buildings or fences are injured, they will be repaired and charged to the occupant.

No one will be allowed to keep swine.

*The 'Merrimack' was an offshoot of the original 'Boston'.

Lunch-hour was half an hour; the dinner-hour was forty-five minutes. 'That this time is too short with due regard to health must be obvious to all. And yet it is probably as long as most business men allow themselves.'[20]

Girls, who were contracted for one year, worked in buildings that were models of brightness. They thrived on the strict regime. A letter:

Dear Friend,

I take my pen in hand to write to you to let you know that I am a factory girl and wish you was one. You dont know how pleasant it is here. We can see everything from our window. There is a lot of hansome fellows.

Nathaniel Hawthorne regarded the girls as 'the bluebells in fashion's nosegay', and Charles Dickens was equally impressed:

I happened to arrive at the first factory just as the dinner hour was over, and the girls were returning to their work; indeed the stairs of the mills were thronged with them as I ascended. They were all well dressed, but not to my thinking above their condition, for I like to see the humbler classes of society careful of their dress and appearance . . . they were healthy in their appearance, many of them remarkably so, and had the manners and deportment of young women: not of degraded brutes of burden. . . . I cannot recall or separate one young face that gave me a painful impression; not one young girl, whom assuming it to be a matter of necessity that she should gain her daily bread by the labour of her hands, I would have removed from those works if I had had the power.[24]

Instruction was provided in the 'domestic arts' and lectures were given by visiting professors, including John Quincy Adams and Ralph Waldo Emerson.

[And] they have got up among themselves a periodical called 'The Lowell Offering' a repository of original articles, written exclusively by females actively employed in the mills . . . [which] will compare advantageously with many English annuals.[24]

The indifference towards education in America had disappeared. Literacy was certainly much higher than in England. An English visitor noted:

The humblest labourer can indulge in the luxury of his daily paper, everybody reads, and thought and intelligence penetrate through the lowest grades of society.

Not, then, just at Lowell, as this was New York. But when Dickens first visited Lowell in 1842, the first fresh bloom of idealism had gone. He was given a specially conducted tour: was it one

of the first PR jobs in American history? Dickens was kept unaware that 'the printed regulations forbade us to bring books into
the mill'[25] and 'the overseer caring more for law than gospel
confiscated all he found. He had his desk full of Bibles.'[25]

Lucy Larcom, mill girl, also objected to visiting clergymen
urging her to give freely from her twenty-five cents a day to the
pioneer church in the West, and what she calls her 'poemlet', in
the 'Lowell Offering', has the air of subversive literature.

THE COMPLAINT
OF A NOBODY
by Lucy Larcom
aged 13.
In which I compare myself
to a weed growing up in a garden.

When the fierce storms are raging
I will not repine.
Though I'm heedlessly crushed in the strife
For surely 'twere better oblivion were mine
Than a worthless, inglorious life.

She commented later:

Now I do not suppose that I really considered myself a weed, though
sometimes I did fancy that a different kind of cultivation would tend
to make me a more *useful* plant.[25]

Already there is a hint that education breeds dissatisfaction with
factory routine. Ellen Collins, another mill girl, wrote in 1840:

Up before day at the clang of a bell. Out to the mill by the clang of a
bell. Into the mill in obedience to the ding-dong of a bell—just as
though we were so many living machines.

Utopia couldn't long survive the profit motive. Many of the girls
were up in arms over their lack of rights. Sarah Bagley, an opera-

*'The bluebells in fashion's nosegay' were
taught in their spare time how to sew*

31

Elias Howe, 1819–1867

tive in 1845, talks of 'these drivelling cotton lords', 'this mushroom aristocracy' who 'aspire to lord it over God's heritage'. Nevertheless, Charles Weld, from England, was still very impressed by the 'young ladies', though he was disappointed to find that the 'famous Lowell Offering Periodical has been discontinued'.[26]

By 1850, half the Lowell operatives were Irish immigrants brought in deliberately to fill the places of the dissatisfied farmers' daughters. Idealism, always an uneasy bedfellow of Mammon, was finally kicked out. The Three Ms of early American industry —morality, machinery and money-making—were put into reverse order.

Despite the failure of the factory idyll, the new nation had flexed its muscles and solved many of its problems. There was a new confidence. Emerson:

Ours is a country of beginnings, of projects, of vast designs and expectations. It has no past. All has an onward and prospective look.

More popularly and with less eloquence:

The Yankee Nation can beat all Creation.[1]

By 1830 the inventor was elevated to the status of folk hero and, as public expectation of material well-being began to grow, became part of the American myth. There was no doubt that, coupled with a good idea and the will to work, the opportunity in the Land of Opportunity was now the opportunity to make money.

The classic rags-to-riches story is the career of Elias Howe (1819–67), inventor, or one of the inventors, of the sewing machine. His life story has all the ingredients of nineteenth-century melodrama; it was repeated and embellished frequently for the benefit of youngsters who might profit by emulation. It went something like this:[1,3]

Dear Reader. Know that Elias Howe was extremely poor and very lame. As a boy he had run away from his father's farm to the famous factories at Lowell. From there he went to a machine shop in Boston. He earned a meagre nine dollars a week. Little more than a boy himself and with a beautiful wife and three children to feed, he decided to try and invent a sewing machine. He worked long hours into the night.

In order to concentrate on his machine, what Howe actually did was leave his job and live off his father. He was basically lazy, but had a driving ambition to make money.

He began by trying to copy the movements of his wife's hands as she sewed and repaired their clothes. But everything seemed to go wrong.

It all seemed hopeless. He was on the point of giving up when, one day, through long perseverance, he found the answer. A machine that worked.

In 1844, Howe came up with the idea of an eye-pointed needle and the use of two threads: one attached to a needle that moved horizontally, back and forth into the cloth (which hung vertically from a clamp), a second held by a needle that shuttled backwards and forwards underneath the cloth, forming a lock stitch with the loop made by the thread from the first needle. He got the idea of the shuttle from watching weaving machines.

In great excitement, he took his machine to the Quincy Hall Manufacturing Company and challenged them to a race. He would place his machine in competition with five seamstresses and see who could finish the work first. One machine against fifty fingers. The race was fast and exciting. Everyone worked at fever pitch, till, exhausted, Elias Howe won. The manufacturers were very impressed but placed no orders for his machines. Howe was destitute.

Elias Howe in competition with the seamstresses of the Quincy Hall Clothing Manufacturing Company, July 1845

Although the machine did 250 stitches a minute, the reason the bosses didn't buy it was sheer economics. It cost $300. He decided to try his fortune in London, and before leaving America in 1846 he patented his machine. He sold the rights for £250 to an English corset-maker called William Thomas, who was also to pay a royalty of £3 for every machine sold. Howe spent eight months adapting his machine to making corsets. And was then fired. By now his wife was ill, and Howe was forced to beg in the streets for money to send her and the children back to America. He was reduced to living in a one-room slum and, to follow his family, had to pawn his letters patent and his last machine.

When Howe finally returned home, he discovered that all his goods had been lost in a shipwreck and that his dear wife lay dying. He hurried to her bedside just in time for her to expire in his arms. The end of the world seemed to have come for Elias Howe. But all was not lost. It came to his notice that whilst he had been away, other men had been using his machine without his permission. Howe decided to fight them in the courts and obtain his just reward.

America had fallen in love with machines. Several which used Howe's principles had been designed in his absence. They had been demonstrated at country fairs and church suppers and were now beginning to sell. Bradshaw's lock-stitch machine was already in use by 1848, stitching clothes for sailors.

Howe was victorious. He was honoured by every country in the world and most of all by his own grateful countrymen. He died a very rich, very happy and very successful man.

*Isaac Singer, 'Deserted but Determined',
makes final adjustments to his sewing
machine. The drawing, and the sentiment, is
typical of the aspirations of mid-century
America and was printed in Singer's own
publicity publication 'Genius Rewarded or
The Story of the Sewing Machine'.*

The moral of this story is—obvious.

Howe was not the first man to make a sewing machine. An Englishman, Thomas Saint, had devised one in 1790, but it was never put to practical use. By 1829 a Frenchman, Barthelmy Thimmonier, had produced a workable machine but by the time he got his US Patent in 1850, he'd been superseded by, amongst others, Isaac Merritt Singer (1811–75). Singer was Howe's chief opponent in what the mass-circulation newspapers called 'The Sewing Machine War'. And, although he lost the court case and Howe made a fortune out of royalties, Singer made a mint.

Singer adopted many of the principles of earlier machines but, in his own words:

Instead of the shuttle going round in a circle I would have it moved to and fro in a straight line and in place of the needle bar pushing a curved needle horizontally, I would have a straight needle and make it work up and down.[27]

That innovation meant that the machine could sew any kind of seam and, instead of being operated by hand, could be worked by a treadle, leaving the hands free to guide the cloth. It was the first really practical machine, simple enough for everyone to use, and the basic design of the machines still used today. 'Next to the plough, this sewing machine is perhaps humanity's most blessed instrument.'[27] It made possible mass-produced, ready-made clothes from garment factories and clothes more easily made by the housewife at home—though whether it actually eased the housewife's task, as Singer claimed, is doubtful.

Where is the woman who can say that her sewing is less a tax upon her time and strength than it was before the sewing machine came in? . . .

*'W. S. & C. H. Thomson's Skirt Manu-
factory' as published in 'Harper's Weekly',
1859. The motto reads* STRIVE TO EXCEL.

34

As soon as lovely woman discovers that she can set ten stitches in the time that one used to require, a fury seizes her to put ten time as many stitches in every garment as she formerly did.[28]

Singer's factories were enormous – foundries, japanning furnaces, the screw department, the needle department, rooms for assembling, ornamenting, polishing, testing. A private railway siding, a quay, even a company steamer. Almost as big and certainly as important were Singer's palatial showrooms. Singer was the first man to use mass-production to open up a new and huge domestic market.

In the 1850s, machines retailed at $100. They didn't sell too well. But with Singer's instalment plan at $5 down and $3 a month, they did. By 1874 they were advertising:

A NEW DEPARTURE. The Singer Manufacturing Company propose to sell ALL VARIETIES of Sewing Machines . . . not imitations, but wonderful improvements embodying in three leading styles the latest developments of every principle which has succeeded. WE NOW HAVE WHATEVER YOU WANT. COME AND SEE IT.

Isaac Merritt Singer, 1811–1875

One of the first hints of built-in obsolescence. The concept of updating in order to make the original item outdated. 'Old Machines taken in Exchange.' Singer didn't invent advertising, but he was the first man to spend a million dollars on it. There was the promotional song, with giveaway sheet music, like 'The Merry Singer':

On the Sandwich Islands
If the towns you scan,
Where it takes but little
To clothe a man;
Even there you'll see me,
Hear my merry lay,
As the dusky maidens
Round about me play.

Out in distant China,
Borneo, Jap-an;
Africa where natives
Eat their fellow man;
Dress but to the knee
There you'll find me singing
Sure as Fate's decree.

CHORUS
I'm a Singer, I'm a Singer—
I'm a merry Singer, I'm a Singer
I'm a merry Singer—singing for you!

Elias Howe's original sewing machine, 1845

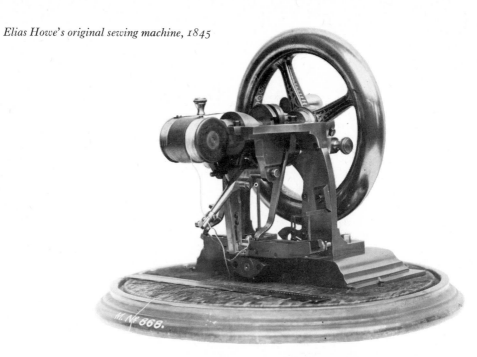

Isaac Singer's original sewing machine, 1854

To my Friend

JOHN Mc. CONNELL, ESQ.

LOUISVILLE, KY.

Song of the Sewing Machine.

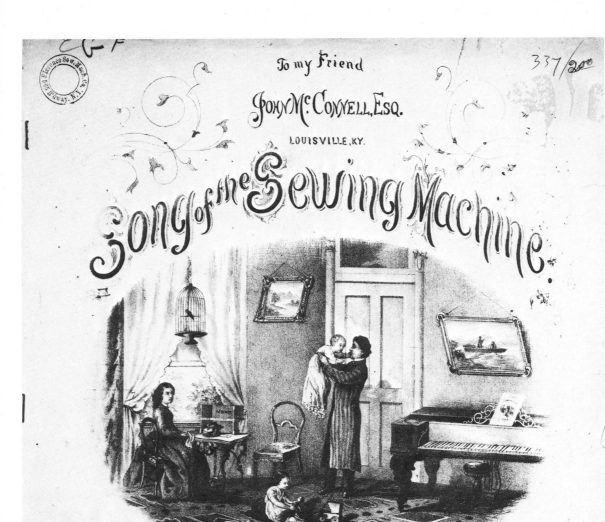

Written & Composed by

WILL. S. HAYS.

THE FOLLOWING SONGS BY Mr HAYS ARE HIGHLY RECOMMENDED.—THEY ARE ALL POPULAR.

Coraline.—Down by the deep Sad Sea.—Mistress Junks.—Driven from Home.—Gay young Clerk.—Good Bye my Boy.—We parted by the River Side.—
Good Bye Old Home.—You've been a Friend to me.—I'm still a friend to you.—Shamus O'Brien.—My Southern Sunny Home.—Nora O'Neal.—
Beautiful Girl of the South.—Katy McFerran.—Kiss me good bye Darling.—Prettiest Girl I know.—Little Sam.—Kitty Ray.—Darling Linnie Dorn.—Jessie Dean.—
. . . . . Kiss the Baby.—Moon is out to night.—Write me a letter.—Laura Lee.—Old Uncle Ben.—Jessie.—No Name.—Belle Bradley.—Darling Kate.— etc. etc.

EHRGOTT, FORBRIGER & CO., LITH., CINCINNATI.

NEW YORK,
JOHN L. PETERS.

ST. LOUIS,
J. L. PETERS & CO.

NEW ORLEANS,
Louis Grunewald.

CINCINNATI,
J. J. Dobmeyer & Co.

GALVESTON,
T. Goggan.

BOSTON,
White, Smith & Perry.

Entered according to Act of Congress in the year 1869 by J.L.Peters in the Clerks Office of the U.S.Dist. Court for the District of N.Jersey.

4

ALL OVER THE WORLD

SINGER SEWING MACHINES

Also Wheeler & Wilson Sewing Machines
at Singer Stores
in your own City

*The American method of manufacturing—
The Singer Sewing Machine Factory,
1853. The machine operation is in the centre
of the floor, with hand-finishing at benches
on either side.*

Under the guise of philanthropy, Singer pushed up sales at home and abroad.

American machines, American brains and American money are bringing the women of the world into one universal kinship and sisterhood.[27]

Advertising in the US amazed European visitors. The French composer Offenbach noted in a parade in 1876 the bass drum advertising a drug store. A mustard firm carved its name and address in the pavement (Americans read sidewalk); telegraph poles, a kilometre apart, read ONLY—CURE—FOR—RHEUMA-TISM.

The range of advertising emphasised the range of goods on the market. In the 1840s the US Patent Office issued 5942 new patents. In the 1850s, 23,140. The value of US manufactures passed the billion-dollar mark and, in the next decade, doubled again.

In 1851 the British Government, worried by the extent and quality of American goods on display at the Great Exhibition at the Crystal Palace, sent a commission to America to find out how they were made. Messrs Wallis and Whitworth reported in 1852:

The American method of Manufacturing Principle, is the production of large numbers of standardised articles produced on the basis of repetition in factories—whereas in England handicrafts and work outside the factories still persist.[29]

They added:

[We] could not fail to be impressed from all that [we] saw there, with the extraordinary energy of the people, and their peculiar aptitude in availing themselves to the utmost of the immense natural resources of their country.

As a result:

The commission at once ordered a large quantity of American gun-making machinery and secured the services of American mechanicians to accompany it to England, and there establish the Enfield rifle fac-

The Centennial Exhibition, 1876—a small section of the half-mile of sewing machines

tory, the first in Europe in which arms were made on the American interchangeable system.[30]

It was the complete reverse of the state of affairs a mere fifty years before.

The Americans in the nineteenth century were like the Japanese in the twentieth—forced to borrow other people's ideas, improve them, adapt them to their own needs and sell them back to their originators. At America's own 'Great Exhibition', the Philadelphia Centennial Exhibition of 1876, America gave notice to the rest of the world that the baby in the New England industrial cradle had grown into a giant. On display were half a mile of sewing machines.

Chapter 2

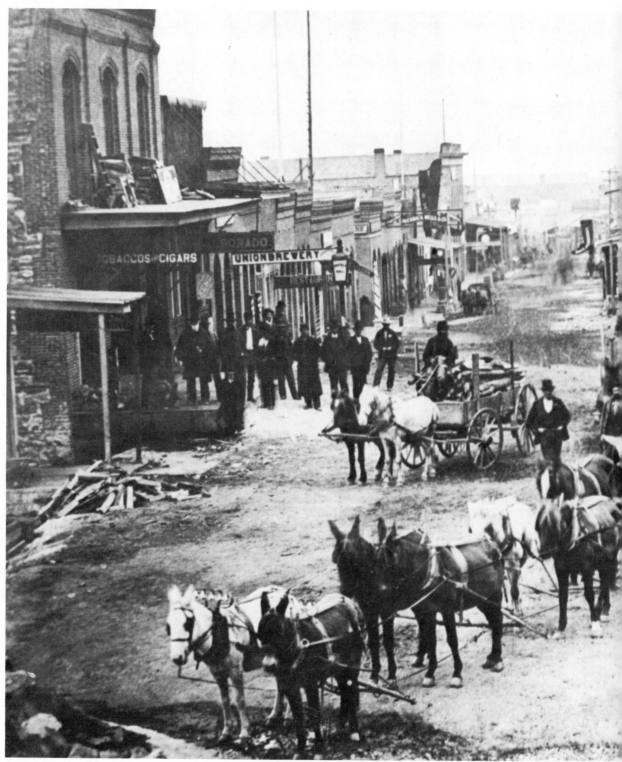

Main Street, Helena, Montana, 1874

Opening up the West

God has predestined. Mankind expects great things from our race and great things we feel in our souls. The rest of the nations must soon be in our rear. We are the pioneers of the world; the advance guard, sent on through the Wilderness of untried things, to break a path in the New World which is ours. In our youth is our strength; in our inexperience is our wisdom.

Not all the inhabitants of North America shared Herman Melville's 1850 view. Like Big Eagle of the Santee Sioux:

The whites were always trying to make the Indians give up their life and live like white men—go to farming, work hard and do as they did—and the Indians did not know how to do that, and did not want to know anyway—If the Indians had tried to make the whites live like them, the whites would have resisted, and it was the same way with many Indians.[32]

The Wilderness, the great unknown, began to yield and become less frightening as the Pioneers turned west. As early as 1769 Daniel Boone went west across the headwaters of the Tennessee River, through the Cumberland Gap, and on northwards through rolling forests and open prairies to the blue-grass country of Kentucky. He was harassed by Indians but the tales his party brought back stirred the imagination.

Horses stained to the knees with the juice of wild strawberries; grapes so plentiful that the quickest way to pick them was to chop down trees; pigeon roosts a thousand acres in size; wild turkies so fat that their skins split open when they were shot to the ground; deer, elk and buffalo in fantastic numbers.

41

'Captain Lewis and Clark holding a Council with the Indians', 'Captain Clark and his men shooting Bears', 'Captain Lewis shooting an Indian'—illustrations from 'Journal of Voyages' by Peter Crass, published in 1812, the first account of the Lewis and Clark expedition to appear in print

A little truth, much exaggeration—already the American love of hyperbole, but sufficiently convincing for 26,000 people to tramp west to the Cumberland settlements of north-central Tennessee during the summer months of 1795.

In 1803, the US purchased the city of New Orleans and an uncharted Louisiana, 880,000 square miles of mid-America, for 15 million dollars. Approximately four cents an acre. The following year, President Jefferson briefed Captain Merriwether Lewis and Captain William Clark for what turned out to be the most significant transcontinental expedition in American history.

The object of your mission is to explore the Missouri river, & such principal stream of it, as, by its course & communication with the waters of the Pacific Ocean, offer the most direct & practicable water communication across the continent, for the purposes of commerce.[33]

The expedition travelled through the West for three years, seeing strange animals, birds, trees, plants, Indians—for the most part friendly—and a country that even the most optimistic New Englander could hardly have imagined. In a long letter to his mother in 1805, Lewis reported that the prairies are not 'barren, sterile, sandy', but that the vast Missouri river 'waters one of the fairest portions of the globe. Nor do I believe that there is in the universe a similar extent of land equally fertile.'

In the steps of the explorers came the speculators. An advertisement for Scioto County, Ohio, placed in French newspapers in 1787, promised 'the most salubrious, the most advantageous, the most fertile' country on earth. It also lied that 500 Americans were already erecting houses and cultivating the 'garden of the universe' for any settlers lucky enough to buy land.

Spurred on by hope, ambition, government decree and these entrepreneurial blandishments, the East-coasters, swelled by the influx of new immigrants, began to turn west, to turn pioneer. Life was hard, but within two decades the West started to provide food to feed the growing East-coast population, and raw materials to feed their growing factories—iron, coal, gold, oil and, later, potash, zinc, copper. But before that could happen the country, wild with animals and Indians, had to be subdued. Guns played a major part.

The one man who didn't make a fortune out of the sewing machine was Walter Hunt—the safety-pin man. During the sewing-machine war of the 1850s Hunt, under oath, had testified that:

Elias Howe has several times stated to me that he was satified that I was the first inventor of the machine for sewing a seam by means of the eye-pointed needle, the shuttle and two threads, but said that he had the prior right to the invention because of my delay in applying for letters-patent.[27]

The reason Hunt had delayed was on the moral grounds that the use of his invention would throw too many seamstresses out of work. No such moral qualms delayed his application to patent his gun for shooting Indians. His, and America's, philosophy was called 'Manifest Destiny'. Thomas Jefferson to James Monroe:

It is impossible not to look forward to distant times when our rapid multiplication will expand itself beyond these limits and cover the whole northern, if not southern continent, with a people speaking the same language, governed in similar forms and by similar laws: nor can we contemplate with satisfaction either blot or mixture on that surface.[34]

One blot on the surface was the black man, another was the Indian. By the 1840s the prediction was being fulfilled.

New territory is spread out for us to subdue and fertilise; new races are presented for us to civilize, educate and absorb; new triumphs for us to achieve for the cause of freedom. [Senator Dickinson][34]

Time . . . will assert and maintain our right with resistless force . . . our population is rolling towards the shores of the Pacific with an impetus greater than we realise. [Senator Calhoun][34]

And, for good measure, 'the call is from Heaven'.

In 1863 a Minnesota General Order encouraged the killing of Indians by offering to all bounty hunters up to $100 per scalp. The order, like the attitude, was typical. The Indians, like the animals, were merely part of the hostile environment that the white man had to overcome. 'Manifest Destiny', genuinely and sincerely believed, was also a neat justification for colonial expansion and the barbarity that often went with it. At that time, the only Americans to object to the killing were the New Englanders. They'd got rid of their Indians already.

The most frequently used gun in the early trapper/settler days was the Pennsylvania-Kentucky rifle, developed from German designs and nicknamed the squirrel gun. Accurate for hitting small moving animals, it was deadly for big moving British Red Coats and it 'made possible the settlement of a continent, freed our country of foreign domination'.[35]

The gun's long barrel had a rifled bore—that is, the inside of the barrel had spiral grooves which meant that the bullet, spinning out of the gun, flew faster and truer than with the older smooth-bore musket. Another, even more important, feature was the speed with which the gun could be loaded. Instead of every

Samuel Colt and the original Colt revolver as patented in 1835. The patent stressed 'more particularly the central fire ignition than the ratchet motions for rotating the chamber'. Another innovation was the introduction of a conical bullet in preference to the spherical one then in use.

hunter/soldier having to carry an iron rod and a hammer to jam the bullet down the gun barrel—cumbersome and time-consuming—the bullet was cocooned in greased cloth, slid easily into the barrel, and was poked down with a wooden stick. It meant that, in battle against Indians, rifles could be loaded on the run. But a man couldn't easily reload on horseback, and the wars against the Plains Indians and the Indians of the South-west were increasingly mounted wars. Stop to reload and a man was a dead duck. The answer was the six-shooter.

Samuel Colt (1814–62) didn't invent the repeating rifle but he did perfect it. Basically, what he did was replace the one-shot shotgun with a gun that had a revolving chamber. There was a bullet in each cylinder of the chamber and, as the gun's firing pin discharged a shot, a ratchet rotated the chamber, locking it into position so that another bullet was ready to be fired at once. Colt was granted his letters Patent in 1835 and 1836.

The early guns were mass-produced at Eli Whitney's factory at Whitneyville, where special machines were devised to make the new and intricate interchangeable parts.

By Consent of the Mayor
An Exhibition of
Colt's Patent
REPEATING RIFLES
will be made at the Battery
On Monday afternoon, 19th inst.
At half past 4 o'clock.
These rifles are eight times more effective
and very little more expensive
than the ordinary Rifle of equal finish.[36]

The date was 1838, the price $150 each, and the success immediate.

In the late Indian fight, Captain Andrews used one of Colt's Patent Rifles which he discharged ten times whilst a comrade could discharge his rifle only twice. I think these rifles in proper hands will prove the most useful of all weapons in Indian warfare. [*The Telegraph and Texas Register*][38]

The prophecy proved right. An early version of the 'unsolicited testimonial' was written to Colt by Captain Samuel Hamilton Walker, Captain of the Texas Rangers in 1846.

Dear Sir,
The pistols which you made for the Texas Navy have been in use by the Rangers for three years, and I can say with confidence that it is the only good improvement that I have seen. The Texans who have

learned their value by practical experience, their confidence in them is so unbounded, so much so that they are willing to engage four times their number. In the Summer of 1844 Col. J. C. Hays with 15 men fought about 80 Comanche Indians, boldly attacking them upon their own ground, killing & wounding about half their number.[36]

The Indians saw things differently. Little Crow, chief of the Santee Sioux, 1862:

See the white men are locusts when they fly so thick that the whole sky is a snowstorm. You may kill one, two, ten—and ten times ten will come to kill you. Count your fingers all day long and white men with guns will come faster than you can count.[32]

But even for the white man there were still unexpected hazards:

Austin, Texas
November 1840

Dear Sir,

A young man named Hotchkiss was dangerously wounded on the 17th by the accidental discharge of one of Colt's Patent Rifles. The inhabitants, by way of ridicule, call the weapon 'Colt's Patent Wheel of Misfortune'.[38]

45

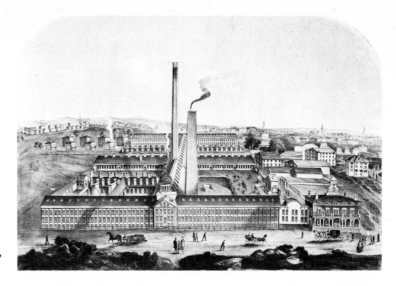

Colt's Patent Fire Arms Manufactory, Hartford, Connecticut

Especially in the early days the guns could backfire—literally, if two charges exploded at once. But, by 1870, the Colt revolver, along with other firearms like the Winchester, Springfield and Sharpes, had reduced the native population of the United States by half. But its subsequent notoriety through the novel and movie has tended to overshadow the fact that the Wild West was originally the Ohio and Indiana of the 1800s, and what opened it up was not the gun but the plough.

The speed of that opening up was tremendous. The change from Wilderness to corn field, from primitive encampment to comfortable community could and did happen in the lifetime of one man. William Conner, for example, born in 1777.

The banks of the White River in Indiana Territory were lined with 'one great forest so dense you couldn't see the sun and sky'. In small clearings were the 'towns' of the Delaware Indians who had settled there, dispossessed of their lands in the East. William Conner, working as a fur trader in and out of Indiana from a base in Michigan, received his first trader's licence from William Henry Hamilton, governor of the Territory, in 1801. He was one of just half a dozen other white traders licensed that year and working in the region of the Delawares. He settled near a small, rare, treeless meadow in a bend of the White River.

It was a beautiful moonlight night in August 1802. With only the help of a French Canadian, I finished the roof on my cabin. It seemed to me, in this Wilderness, a major achievement. It made me very happy.[39]

The 'double' log cabin consisted of two rooms. It had an earth floor, bunk beds made out of sassafras wood to discourage insects, windows made of paper or skin rubbed with bear grease to make them waterproof, and a stone fireplace. Conner lived in his cabin for upwards of eighteen years. For food he had shelled corn grown in the cleared patch at the back of the hut, and venison

shot and dried in summer to eat in winter. For clothes he had trousers, jerkin and cap made from animal pelts or tanned leather. He traded with the Indians—beads, knives, blankets, gun flints, tobacco, salt, spirits, traps, corn and tools like 'trade axes'. In return, he received skins—bear, fox, cat and mink. He spoke Delaware, thought of himself as Delaware, and sometime after 1802 married Mekinges, a Delaware girl, and fathered six children. They lived together in the cabin-cum-trading post and, over the years, as well as serving as Indian scout for the territorial governor, taking part in two battles and representing the Delaware in their dealings with the government, Conner worked the land in the bend of the White River and planted it with corn.

In 1816 Indiana became a state, and in 1818 the Delaware nation signed a treaty that ceded their claims to all Indian lands which, along with the land ceded at the same time by the Wanydotte, Seneca, Shawnee, Ottawa, Pottawatomie, Miami, Wea and Chippewa nations, amounted to about two-thirds of the northern part of the state. In exchange, they were to get the cash value of the land plus $4000 annually, a blacksmith and the guarantee of peaceful possession of 'a country to reside in upon the West side of the Mississipi'. It was to be another forty years before Motovato, Black Kettle of the Southern Cheyennes, was to say, 'It is hard for me to believe the promises of white men any more'.[32]

In the early summer of 1820, Conner stood outside his cabin:

Hé knew now, as a certainty, that Mekinges and the children would leave for the west with the Delawares in obedience to the tribal law that the Indian wife must stay with her people . . . her ways were not the ways of white women. . . . The proud old chief took his place at the head of the procession and the trek westward began. William Conner gazed until they were lost to view and the last bit of dust raised by the ponies' feet had sunk to the ground. He reflected proudly that Mekinges was the best dressed of all the Indian women.[39]

47

Although the immediate cause of the Delaware departure was the new treaty, the real cause was the pressure of white immigration. As early as 1805, people began streaming into southern Indiana—not along the trails which were difficult to negotiate, but in flat-bottomed boats, 'floating down the River on the O-H-I-O'.

Within a decade, the trickle had become a flood. Half a million Irish, English, Scots and Germans moved west. Travelling in covered wagons along new routes, one group in bad weather took ten days over a journey of nineteen miles. But in good weather, and on foot, settlers could cover ten miles a day, and on horseback forty. Government land sales in 1833 totalled four million acres. By 1836 it was twenty million. The frontier moved forward an unstoppable seventeen miles a year. Conner suddenly had neighbours and he crossed back into the white world.

In 1820, he married Elizabeth Chapman, seventeen-year-old daughter of a newly settled family. He bought some land and built himself one of the first brick mansions in the 'New Purchase'—the old Indian lands. 'The delicate mantels, stairways and glass-door cupboards . . . were light and unusually delicate for this region.'[39] The house overlooked the land in the horseshoe bend of the White River. It was, and is, 'remarkable handsome'.

As well as a new home, the house served as post office and court house for the new Hamilton County, founded in the same year, 1823. And Conner himself later became a representative in the Indiana General Assembly. In many ways, he was the archetype of Emerson's pioneer yankee, 'in turn trapper, farmer, congressman, jack of all trades'. And he lived to see his land transformed into the bread-basket of the community. As a visitor in 1823 noted:

William Conner and his 'remarkable handsome' brick mansion—one of the first to be built in the old Indian lands

48

I never beheld a scene more delightful than when I look down on that field. Three hundred acres of waving corn, some two feet high. And those fifteen or twenty men scattered over it—all at work. It was doubly interesting coming, as it did, out of nature's forest, only broken by the occasional cabins and the small patches of cleared land of the early settlers.[39]

The 'small patches' across the West were not to remain small for very long, and their growth, like the triumph of Conner's three hundred acres of waving corn, was the achievement of white sweat and white technology—a force with which the Indian could not and would not compete.

The traditional American ploughs (Americans read plow) were based on European designs and modified to suit local needs. It was the ubiquitous and versatile Thomas Jefferson who, in his study of the plough, made the important discovery that the throat or cutting edges of the plough had to be straight. In 1797 Charles Newbold replaced the wooden ploughshares with iron, and in 1814 Jethrow Wood created the first plough with interchangeable parts. At first, the iron ploughshare didn't find acceptance as farmers thought that iron would poison the earth, and Newbold was penurised. But, slowly, iron was accepted and these New England ploughs proved themselves very useful for cutting through the light soils of the East. In the thick, heavy soils of the West they were useless. They refused to scour—to make the soil fold back and away from the mouldboard to cut a clean, deep furrow. 'In the very nature of the soil lay an embargo that seemed almost unsurmountable.' Many farmers were in despair: 'Pass the prairies and head for the hills. . . .' The soil was lighter there.

John Deere, 1804–1886

John Deere of Vermont moved to Grand Detour, Illinois, in 1836. He was a blacksmith and repaired the young community's broken ploughs, forging the iron ones and reinforcing the wooden ones with strips of iron or steel. But Deere was another man with vision. He began to experiment with different shapes and materials for plough bottoms in an attempt to solve the problem of the heavy soil. A prototype plough, with a share made from a piece of circular steel saw, was ready for field trial in the spring of 1837. Mr Lewis Crandell's farm in Ogle County had the reputation as the stickiest black soil in Illinois. But, as Deere steered his new plough, creating furrow after furrow, the black earth fell smoothly away. Steel was self-scouring: the steel surface and the curved shape of the mouldboard were the answer to the farmer's prayer. In the years that followed, John Deere's 'Self-Polisher' cut a furrow deep into the Indian lands.

John Deere's first steel plough built in Grand Detour, Illinois, 1837, and demonstrated on Lewis Crandell's farm, Ogle County, Illinois

John Deere's first plough factory at Moline, Illinois

Cyrus Hall McCormick's reaper, invented on Walnut Grove Farm, Virginia, in 1831. The wide master wheel carried most of the machine's weight and through ground traction supplied power to operate the reel and the reciprocal knife. Several men were still required to bind up the grain but compared with the old method of scything with 'cradles' (above), harvesting could be completed in under half the time with one quarter the manpower.

The McCormick Reaper Works—situated on the north bank of the Chicago River and burnt down in the great Chicago fire of 1871

51

Cyrus Hall McCormick, 1809–1884

At the same time, twenty-two-year-old Cyrus Hall McCormick was perfecting the reaper, the biggest advancement of all in the mechanisation of American agriculture. In the days of William Conner, a man working a field with a scythe averaged an acre a day. What wasn't cut in two weeks of good weather at harvest time was left to stand. Farmers were very careful not to plant more than they knew they could gather, and to make up the shortfall in the American wheat crop grain was imported from Europe.

But in good years wheat was also exported. In 1839 Chicago shipped eighty bushels, in 1849 two million bushels. The reaper made it possible for farmers to plant ten, twenty, even thirty times the amount of wheat and still be sure of harvesting it. The reaper made the West pay.

The McCormick reaper, horse-drawn like the plough, was developed in the Shenandoah Valley of Virginia in 1831. There had been attempts at making grain-cutting machines before. What made this one different was that it worked. It wasn't so much the product of great imagination as of clever technological organisation.

McCormick's approach was to break down the harvesting process into separate actions and analyse and overcome each problem individually. The first problem was how to divide the wheat to be cut from the wheat left standing. Answer: a divider—a curved arm to bend the wheat towards the knife. Next, how to make sure all the stalks of wheat are cut? Answer: a reciprocating blade, a knife that moved backwards and forwards. What the blade failed to cut with one movement it chopped on the other. Problem three: how to keep the wheat standing so that the knife didn't merely flatten it? This was solved by attaching metal 'fingers' to the blade so that the wheat would be caught and held in position while the knife cut. Any low or fallen stalks were lifted up and pressed into the fingers by a device called a 'revolving reel'. A platform collected the cut wheat, which was then raked off and tied in bundles by hand. So that the horses wouldn't trample over wheat still to be harvested, the shafts of the reaper were placed to one side. And the whole machine was driven by one large 'driving wheel' that operated both the wheel and the knife. McCormick was, in the words of his later rival C. W. Marsh, the inventor of the Marsh Harvester, a man who combined the ideas of other men with his own and 'produced the first practical side-delivery machine in the market'.[40]

In 1832, a public demonstration of the machine was given at Lexington, eighteen miles from the McCormick farm. Professor Bradshaw of the Lexington Female Academy deliberated on what he saw and finally pronounced: 'This machine is worth—a hundred thousand dollars.' The locals, though, weren't so sure.

It had to be led by a couple of darkies because the horses were scared to death by the racket of the machine.

And:

I thought it was a right smart curious sort of thing, but that it wouldn't come to much.[40]

It didn't—at first. The small hilly fields of Virginia were not suitable for it. Not until 1841 did McCormick sell a machine. In 1842 he sold seven. Then he moved West and saw that the flat, rolling prairies were as suitable for his machine as the area was in need of it. More than that, the British harvest of 1845 had failed, and the following year Britain changed her import laws and was crying out for wheat. McCormick seized his chance. He chose the small town of Chicago to build his factory. In five years, he sold five thousand machines, some of them for export.

The Times of London saw the machine at the Great Exhibition of 1851:

A cross between a flying machine, a wheel barrow and a chariot. . . . An extravagant Yankee contrivance, huge, unwieldy, unsightly and incomprehensible.

But, in a special demonstration, it 'mowed down British prejudice and opened the way for the bringing of our countrymen and their contribution before the public in a proper light'.[42]

The Times changed its tune:

The reaping machine from the United States is the most valuable contribution from abroad to the stock of our previous knowledge. . . . It is worth the whole cost of the Exposition. [9 June 1851]

It was awarded the Grand Medal and the Council Medal. Prizes also went to American ploughs, the Colt revolver, a sewing machine based on Howe's principle, Goodyear rubber products, Borden meat biscuits and American-made locks. The London *Evening Chronicle*:

The sceptre is fast passing from England; westward the Star of Empire takes its way.

Times have not changed. All the agricultural machinery in the 1974 procession of the London Lord Mayor was made by the John Deere Company.

The plough and the reaper made all the difference to the new arrivals in the Midwest. No longer was it frontier country.

McCORMICK'S PATENT VIRGINIA REAPER.

D. W. BROWN,
OF ASHLAND, OHIO,

Having been duly appointed Agent for the sale of the above valuable labor-saving machine (manufactured by C. H. McCormick & Co., in Chicago, Ill.,) for the Counties of Seneca, Sandusky, Erie, Huron, Richland, Ashland and Wayne, would respectfully inform the farmers of those counties, that he is prepared to furnish them with the above Reapers on very liberal terms.

The Wheat portions of the above territory will be visited, and the Agent will be ready to give any information relative to said Reaper, by addressing him at Ashland, Ashland County, Ohio.

Ashland, March, **1850**.

From Richard and Margaret Pugh,
Praerie Villa. November 18th. 1845.

To the Rev. Eli Jones,
Minister of the Calvanistic Methodist Chapel,
Aberystwyth, Wales.

Dear Eli,

. . . I found the country better than I expected and have bought an improved farm with house, buildings and eighteen acres of wheat ready sown. The land appears rich and fruitful and I am happy to say that I feel quite at home. . . . I would not for a considerable sum return to Wales. The tax gatherer only calls once a year and then it is only a trifle. This is the country for a man with a family, where provisions are cheap and of the best kind.

It was forty years since William Conner had first come to Indiana.

Other agricultural machines, mowers, threshers, other harvesters—particularly Marsh's of 1858—were quickly on the market. In 1860, over eighty thousand reapers alone were working in the Midwest, and the nation's wheat harvest approached two hundred million bushels. In 1871, Sylvanus Locke produced a wire binder that took over the manual operation with twine. Soon, in the 1880s, came the traction engine and the combine harvester. Already, by mid-century, agriculture was an industry as mechanised as the industries in the Eastern factories. Agrarian mass-production was a reality.

As men poured West, the railroads followed. American engineering had already had its triumphs. After the Revolution, communications had been gradually improved by a network of roads and canals, but the railroad was by far the greatest advance in the technology of transport.

In 1804, Oliver Evans designed a new, high-pressure steam engine which he used to grind corn and saw stone. By 1811, the engine had a horse-power of eleven. Before he died in 1819, Evans made a prediction:

The time will come when people will travel in stages [stage coaches] moved by steam engines, from one city to another, almost as fast as birds fly, fifteen to twenty miles an hour.

Passing through the air with such velocity, changing the scene in such rapid succession, will be most exhilarating, delightful exercise.

A carriage will set out from Washington in the morning, the passenger will breakfast at Baltimore, dine at Philadelphia and sup at New York on the same day.

To accomplish this, two sets of railways will be laid, so nearly level as not in any place to deviate more than two degrees from a horizontal line, made of wood or iron or paths of broken stone or gravel, with a

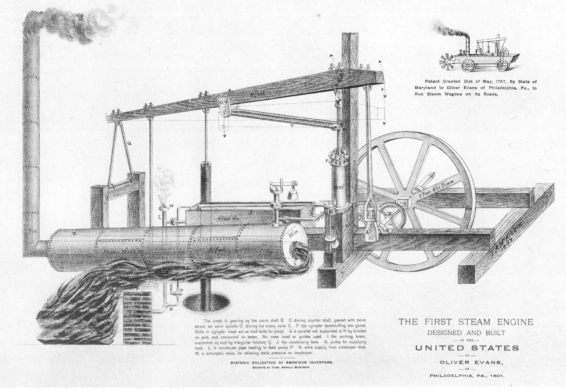

The crank A, gearing on the crank shaft B. C driving counter shaft, geared with bevel wheel, on valve spindle D, driving the rotary valve E. F the cylinder headstuffing box gland. Bolts in cylinder head act as stud bolts for gland. G is parallel rod, supported at H by bracket on post, and connected on beam. No cross head or guides used I the working beam, supported on end by triangular fulcrum Q. J. the condensing tank. K, pump for supplying tank. L is condenser pipe leading to feed pump P. N, extra supply from condenser tank. M, is automatic valve, for relieving back pressure on condenser.

HISTORIC COLLECTION OF AMERICAN INVENTORS.
DRAWING BY THOS. ARNOLD McKINNIN.

THE FIRST STEAM ENGINE
DESIGNED AND BUILT
— IN THE —
UNITED STATES
— BY —
OLIVER EVANS,
— OF —
PHILADELPHIA, PA., 1801.

The first steam engine designed and built in the United States by Oliver Evans of Philadelphia, Pennsylvania, 1801

rail to guide the carriage, so that they may pass each other in different directions and travel by night as well as day. . . .

And it shall come to pass that the memory of those sordid and wicked wretches who oppose such improvements will be execrated by every good man as they ought to be now.[43]

The first US railway was on Boston's Beacon Hill in 1795 and was used to lower bricks from a kiln to the road. The second, in Pennsylvania in 1810, was used for hauling stone. But the first railroad proper was the Baltimore and Ohio, begun in 1828. By May 1830 there were thirteen miles of it and, as Evans had predicted, two tracks. In the same year, the Mohawk and Hudson began construction in New York state and, in 1831, the Charleston and Hamburg in South Carolina opened, carrying cotton and, later, some passengers. Three railroads opened in the surrounds of Boston, one of them to the factories at Lowell. In England, the Stockton and Darlington Railway had opened in 1825.

The earliest and most important technological innovation on the all-American railroads came in the construction of the track itself. In 1830, Robert Stevens went to England to buy wooden rails for the projected Camden and Amboy line. The shape of the British rail meant that it had to be set in an iron chair, and the chair then attached to the wooden ties of the track. To avoid the excessive expense of iron, Stevens invented the broad-based T

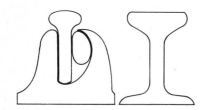

Profile of the British rail set in its complicated iron chair. In contrast, the American rail with its broad base was simply held in place on the ties by long hook-headed spikes.

rail, which did away with the need for a chair and could be attached direct to the wooden ties (British read sleepers) by another Stevens invention, the spike, in effect a long, square-headed nail which hooked over the broad base of the T rail and was driven directly into the tie. It meant that the Americans were able to lay something like twice the amount of track at twice the speed and half the price of the British. Within ten years, Stevens's track-laying method was adopted throughout the world.

Railroads were not originally designed for steam engines. As a Baltimore schoolbook of 1837 put it:

You may mount a car, something like a stage, and be drawn along by two horses at the rate of twelve miles an hour.

Or, if you lived some places in Pennsylvania, by oxen. But all that was changed when, in England in 1829, George Stephenson won an open competition for steam engines by hauling a 13-ton train at 15 mph on the Liverpool-Manchester Railway. His locomotive was 'The Rocket'. As a result of this success, Stephenson was invited to build 'The America' for the Delaware and Hudson Canal Company, a forward-looking outfit who saw the days of the canal were numbered and had set about diversifying. The same company ordered from Stourbridge, near Birmingham, the first engine to run a commercial service in the US. The 'Stourbridge Lion' was a beautiful piece of precision engineering. She ran superbly but, for American rails, was too heavy. She was nicknamed the 'track-smasher' and abandoned. As with other machinery from Europe, America had to modify the design and build her own locomotives to suit her own particular needs.

The first all-American locomotive was the Baltimore and Ohio's 'Tom Thumb' (1830). Weight, one ton; hp $1\frac{1}{2}$; load, 24 passengers on single track; average speed 6 mph. In a much-publicised, steam versus horse-power 'thrilling trial', the 'Tom Thumb' raced a horse car—and lost. Not because it was slower, but because the engine developed a fault.

'The Best Friend of Charleston', the pride of South Carolina and the first practical American locomotive, weighed $4\frac{1}{2}$ tons, had wooden wheels and iron tyres and, in service, reached a speed of 21 mph. But the fireman, bothered by the constant hiss of steam, held down the safety valve. 'The Best Friend of Charleston' blew up. After she'd been rebuilt, she was hopefully renamed 'The Phoenix', and passengers were protected from further explosions by a barrier car, a flat truck between locomotive and passenger cars, piled high with cotton bales. But still, in 1839, for Captain Maryatt:

The locomotive threw such showers of fire that we were constantly in danger of conflagration. . . . As the evening closed in we were whirled along through a stream of fiery threads—a beautiful, although humble imitation of a comet.[44]

The 6-ton De Witt Clinton, of the Mohawk and Hudson, on its first run between Albany and Schenectady in August 1831, reached a speed of 15 mph. Without passenger cars it reached 40 mph. In the same year, the same company's engineer, John Jervis, rebuilt the English import 'John Bull', rechristened it 'Philadelphia', and introduced a system of bogie wheels:

. . . four wheeled trucks ingeniously mounted on swivel axles, enabling it to whisk round curves at the sight of which an English railway engineer would stand aghast.[45]

Jervis also placed the driving wheels at the rear of the engine and increased their size so that they could gain greater speed and do less damage to the track. Equally clever was the trolley on wheels that Isaac Dripps devised for the Camden and Amboy, and which was attached to and intended to run in front of the train. The forerunner of the cow-catcher, it wasn't intended to catch errant cows but to steady the engine. In 1835, William Harris built the 'George Washington', a grade-climbing engine which, for the first time, caused sceptical British designers to buy American.

The craftsman-built engines and the straight, craftsman-built railways of England were a luxury America could not afford. If there was a hill, the train must go over it. If there was a river or a ravine, it must be bridged by whatever materials were available. Frequently, engines fell off the tracks and bridges collapsed—but it didn't matter. The aims of the railroad builders were the same as the devisers of mass-production—to produce quickly and cheaply for the moment.

Mammon-seeking mortals. . . . In America everything is sacrificed to time, for time is money.[44]

The price of speed, and novelty, was passenger discomfort. The American reaction, from Samuel Breck on a journey from Boston to Providence on 22 July 1835:

Uglier boxes I do not wish to travel in. Some thirty human beings sit cheek by jowl as best they can . . . [smelling] of fish, tar and molasses . . . and all this for the sake of doing very uncomfortably in two days what would be done delightfully in eight or ten.[11]

And the British reaction, from Dickens on his trip to Lowell:

There are no first and second class carriages as there are with us; but there is a gentleman's car and a ladies' car: the main distinction between which is that in the first, everybody smokes; and in the second, nobody does. . . . There is a great deal of jolting, a great deal of noise, a great deal of wall, not much window, a locomotive engine, a shriek and a bell.[24]

Nevertheless, by the 1840s Emerson could write:

The Americans take to this little contrivance, the railroad, as if it were the cradle in which they were born.

By 1850, the Americans possessed the largest locomotives in the world, weighing in at 150 tons and doing 70 mph. And the largest railroad network—a total of 9000 miles, three times the amount in Europe. It linked twenty-five states and was the umbilical cord that kept the Midwest supplied with machinery and manufactured goods, and the East, in return, with food.

The 1850s were a watershed in the nation's history, the point at which America changed from an agrarian to an industrial society. It was a change that not all Americans welcomed. Thoreau, for example, sitting by Walden Pond in New England, eating his woodchuck raw, was led to comment:

To do things 'rail-road' fashion is now the by-word . . . But if we stay at home and mind our business, who will want railroads? We do not ride on the railroad. It rides on us.[46]

But there was to be no looking back. Jessup Scott, newspaperman, 1823:

From all enlightened nations of the old world and from all the old states of our Union . . . multitudes are pouring in which swell in magnitude with every revolving year. . . . The imagination can conceive of nothing more imposingly grand than this march of humanity westward, to enter into possession of times' noblest empire.

In the 1830s, the US population was less than thirteen million. By 1850, it was twenty-three million. During that period, states like Indiana and Illinois doubled their population and tripled their agricultural output. The numbers of pioneers taking the Oregon Trail mushroomed: 'Oregon—Land of Promise and Fable'. The Gold Rush to California between 1848 and 1854

First train into Tottenville, Staten Island, New York, 1860—an 'American' type wood burner locomotive

stimulated interest in the remote regions of the Pacific Coast. Travellers' tales, even discounting the get-rich-quick stories about gold, reported on California in much the same words as the men who had accompanied Boone to Kentucky eighty years before. Giant strawberries, grapes of unbelievable size and abundance, and perpetual summer. And this time, the stories were true.

The railroad was the tangible sign of the nation's progress. In 1850, it paused on the banks of the Mississippi. On the opposite side were the inviolable lands of the Indian—the Permanent Indian Frontier already having been shifted in 1834 to the 95th Meridian, 200 miles west of what had originally been promised.

Make way, I say, for the young American Buffalo—he has not land enough. . . . I tell you, we will give him Oregon for his summer shade and the region of Texas as his winter pasture . . . the mighty Pacific and the turbulent Atlantic shall be his. [New Jersey Congressman, 1844]

It was inevitable that the railroad should follow the people, cross the Great River and play its part in fulfilling America's 'Manifest Destiny'. Grenville Dodge, Surveyor of the Union Pacific Railroad, in 1862:

It is a grand anvil chorus that these sturdy sledges [sledge-hammers] are playing across the plains; it is in triple time, three strokes to the spike. There are ten spikes to the rail, 400 rails to the mile; 1800 miles to San Francisco. 21,000 times are these sledges to be slung. 21 million times to come down with their sharp punctuation.

To the British, who only used railways to link up centres of industry, the American use of the railroad to open up a country was an entirely new concept. James Stirling, a visiting Scotsman, in 1856:

There is nothing in history to compare with this seven league progress of civilization. For the first time, the world can see a highly civilized people quietly spreading itself over a vast untenanted solitude, and at one wave of its wand, converting the wilderness into a cultivated and fruitful region.

As important as the railroads in opening up the West and uniting the nation, was the telegraph. A contemporary jingle:

> Hark! the warning needles click,
> Hither, thither, clear and quick.
> Sing who will the Orphean lyre,
> Ours the wonder working wire.

In 1848 James Marshall announced to the world, 'Boys, by God, I believe I've found a gold mine'. In the two years that followed, over 100,000 men went to California in search of their eldorado. Most of them had their photographs taken—to tell the folks back home.

The Railroad back East—
'The 9:45 Accommodation, Stratford,
Connecticut.' A painting by Edward
Lamson Henry, 1867.

The Railroad pushes West—
Laying the wooden ties of the Union Pacific
Railroad in Wyoming, c. 1867

Construction train on the Union Pacific Railroad at Granite Canyon, 20 miles west of Cheyenne, Wyoming, c. 1868.

'Camp Victory' (below), 28 April 1869, after eight brawny Irishmen and a small army of Chinese had laid more than ten miles of rail to establish an all-time construction record and win a wager of $10,000 for their boss. The job was directed by Construction Superintendent J. H. Strobridge, standing on the flatcar in the foreground. The shadowy figure in the doorway of the construction car (top right) is Mrs Strobridge, the 'Sweetheart of the CP' and the only woman to accompany the construction forces throughout the entire period that the railroad was built.

Joseph Henry (1797–1878)—an early daguerreotype, c. 1850

It was the invention of a reasonably successful portrait painter, a man who had painted Eli Whitney. He was Samuel Morse (1791–1872), then fifty-three years old. Morse's telegraph relied heavily on the work of Joseph Henry (1797–1878), the first 'pure' American scientist since Franklin, who, like Franklin, was a researcher in electricity. It was already known in Europe that electric current flowing through a piece of metal will set up a magnetic field. In 1830, Henry had posed the question whether, if electricity produced magnetism, could magnetism produce electricity? His answer: yes.

If a single wire be passed by a magnetic pole, a current of electricity is induced through it which can be rendered sensible.[1]

But the words were not Henry's. Unknown to him, the Englishman Michael Faraday had produced the same answer a year before and reported the fact to the Royal Institution. Undeterred, Henry applied the results of his electromagnetic experiments:

To the invention of the first electro-magnetic telegraph, in which signals were transmitted by exciting an electro-magnet at a distance by which . . . dots might be made on paper and bells were struck in succession, indicating letters of the alphabet.[16]

Henry sent his signals over a mile of wire looped round his laboratory.

About the same time, 1832, Samuel Morse was on board the *Sully*, sailing home from Europe. In casual conversation about electrical experiments, he was assured by Professor Charles Jackson that it was possible to conduct an electrical impulse over a long wire. Morse saw the implications:

If this be so and the presence of electricity can be made visible in any desired part of the circuit, I see no reason why intelligence might not be instantaneously transmitted by electricity to any distance.[47]

Morse's previous experience with electricity was as an undergraduate at Yale, a quarter of a century before, but he was not to be put off by his comparative lack of scientific background. 'The American Nation', after all, 'can beat all Creation'; so could a single middle-aged American if he applied himself. With his sketch for a primitive telegraph system, Morse's career as a professional painter came to an end.

Henry continued his experiments, in 1836 constructing the first telegraph line using the earth as conductor, and sending signals from his house to his lab. Morse, independently, continued his. His whole apparatus was small and simple: a battery, two sets of wires for transmission in each direction, an operator-controlled mechanism for breaking and restoring the circuit at one end, and at the other a receiving device which held a pencil for recording the electrical impulses on a moving strip of paper.

Unfortunately for Morse, there was already a telegraph system operating in America. It was based on the Frenchman Claude Chappe's 'lightning telegraph', and consisted of a series of poles with arm-like extensions, each in sight of the other and manned by an operator. The position of the arms was varied to spell out simple messages in semaphore and transmit news twelve times as fast as a man could carry it on horseback. The first line built in the US was in 1800 from Martha's Vineyard to Boston. Now, in 1837, Congress was proposing to build a Chappe system to link New York with New Orleans. The problem with the system, though, was that it was useless at night and unreliable in bad weather. Morse pressed forward and presented his ideas to Congress. In 1838, Congress reported:

It is obvious that the influence of this invention . . . will, in the event of success, of itself amount to a revolution unsurpassed in moral grandeur by any discovery that has been made in the arts and sciences . . . space will be, to all practical purposes of information, completely annihilated between the States of the Union.[47]

Proud words—but no action. Morse was down. The man who lifted him was Joseph Henry. When they had first met, Henry thought Morse 'an unassuming and [un]prepossessing gentleman with very little knowledge of the principles of electricity or magnetism'.[47] But, in 1839, he wrote:

I am acquainted with no fact which would lead me to suppose that the project of the electro-magnetic telegraph is impractical; on the contrary, I believe that science is now ripe for the application, and that there are no difficulties in the way but such as ingenuity and enterprise may obviate.[47]

Morse was encouraged to go on. In August 1842:

No one knows the days and months of anxiety and labor I have had in perfecting my telegraphic apparatus. . . . Nothing but the consciousness that I have an invention which is to mark an era in human civilization and which is to contribute to the happiness of millions would have sustained me through so many and such lengthened trials of patience in perfecting it.[47]

Late in 1842, Henry wrote to Morse again:

I am pleased to learn that you have again petitioned Congress in reference to your telegraph, and I most sincerely hope you will succeed in convincing our representatives of the importance of this invention.[47]

Samuel Finley Breese Morse (1791–1872) —self-portrait at the age of 23

The original drawing on which Morse based his patent—battery (A) is linked by wires to a line of ridged pieces of metal (B) fed under a wooden rod (C) that makes and breaks the electric circuit (E) and which is recorded by a pencil (F) marking a continuously moving piece of tape (G)

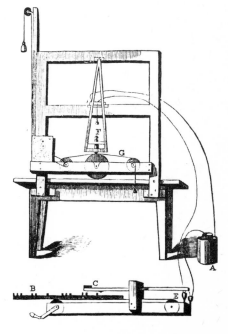

He did. Congress gave him $30,000 to build an experimental line from Baltimore to Washington. The initial problem of not being able to send an impulse along a wire for more than a thousand feet was overcome by Henry. He showed Morse how to introduce a relay system whereby the electric circuit was linked with other circuits at frequent intervals all the way down the line. As a result, the signal was as strong at the end of the line as it was at the beginning. On 24 May 1844, Morse tapped out a message from the Supreme Court in Washington, using his new code 'in which letters, words and phrases . . . were to be indicated by dots and other marks and spaces on paper'.[47] Morse's partner, Alfred Vail, forty-one miles away in Baltimore, transcribed the message from the 23rd verse of the 23rd Chapter of the Book of Numbers: 'What Hath God Wrought'. It was the world's first message sent by electromagnetic telegraph. Morse was fifty-three; it was fourteen years since his conversation with Professor Jackson on the *Sully*, and he had spent seven of them trying to get Congress to take him seriously.

Morse had patented his invention. Henry had not, describing motivations similar to those of Benjamin Franklin one hundred years before:

I did not then consider it compatible with the dignity of science to confine benefits which might be derived from it to the exclusive use of any individual.[50]

The familiar and undignified legal wrangling soon broke out. Morse declared:

While, therefore, I claim to be the first person to propose the use of the electro-magnet for telegraphic purposes and the first to construct a telegraph on the basis of the electro-magnet, yet to Professor Henry is unquestionably due the honor of the discovery of a fact in science which proves the practicability of exciting magnetism through a long coil at a distance, either to deflect a needle or magnetise soft iron.[47]

He made no mention of the practical help that Henry had given him, or that 'had Henry covered his discoveries with patents, all that would have remained of Morse's telegraph would have been the alphabetic code'.[50]

For Henry's part:

I am not aware that Mr Morse made a single original discovery, in electricity, magnetism or electromagnetism, applicable to the invention of the telegraph. I have always considered his merit to consist in combining and applying the discoveries of others in the invention of a particular instrument and process for telegraphic purposes.[16]

which was fair and recognised that the ability to 'combine and apply' was a real and rare talent—witness Singer, McCormick and many other inventors. But Morse was aggrieved:

I am held up by name to the odium of the public. . . . I find the fate of Whitney before me. . . . Take as much pains as you will to secure yourself, and your valuable invention, you will be robbed of it and abused into the bargain. This is the lot of a successful inventor and no precaution, I believe, will save him from it . . . the unprincipled will hate him and detract from his reputation to compass their own contemptible and selfish ends.[47]

Ultimately, Morse, a self-sufficient egoist, had his patent rights upheld. He made his fortune, like his predecessors Whitney, Howe, Deere and McCormick, because his invention, like theirs, coincided with the needs of the nation. Henry, on the other hand, made nothing. He was an idealist, not a businessman.

By 1850, only Florida, of the States east of the Mississippi, was without the telegraph. Despite continual breaks in the lines, despite a construction free-for-all as operators scrambled for profits, order emerged from chaos with the founding of the Western Union in 1856.

The telegraph lines followed the lines of the railroads: they crossed the Mississippi together. Together they revolutionised communications. They were natural allies, and satisfied the two major cravings of mid-century America—speed and efficiency. Businessmen could be wired in advance to expect a particular railroad delivery; they could send and receive important orders faster than by post; learn from the New York Stock Exchange the latest prices of the Western Union, the Chicago and North Western Railroad, and the Union Pacific. Newspapers could splash the latest news across the front pages within hours of the events, and, as early as 1848, wide-eyed readers were digesting the 'News by Lightning Wire'[1] from the war on the Mexican front.

On a personal level, families were able to keep in touch; on an increasingly impersonal one, so were Governments, now able to send out instructions to their representatives anywhere in the country and—after the laying of the transatlantic cable in 1866—anywhere in Europe. Almost from its inception, Morse himself was aware of the power that the telegraph could give to whoever controlled it, to falsify or withhold messages, to manipulate distant situations. The era of centralised government, with its benefits and pitfalls, had begun.

But in the 1850s the more obvious and immediate danger came from the railroad. In 1855, Charles Weld was on a vacation tour.

Telegraph construction kept pace with track laying. Track gangs in the Nevada desert, near Wadsworth, 1868.

'Queen Victoria desires to congratulate President Buchanan upon the successful completion of the great international work, in which the Queen has taken great interest.' The first transatlantic telegraph cable was finally completed on 8 August 1858.

'Interior of a passenger-car at the moment of the collision near Steamburg, New York'

A significant hint of the impending catastrophe was given by the fall of a ponderous lamp-glass on my head but with no worse result than inflicting a smart blow. Presently another glass was jerked from its socket and precipitated into the lap of a lady; the oscillations of the car, meanwhile, increasing in violence.

Affairs now assumed a serious aspect and I felt certain we were on the eve of a smash. This was the opinion of a gentleman who proceeded . . . to instruct us how to place ourselves, laying great stress on the importance of sitting diagonally in order not to receive the shock directly on the knees. We were also advised to hold the back of the seats before us.

In vain was the conductor urged to slacken the excessive speed. With blind, if not wilful recklessness it was maintained and at length . . . a terrific crash and a series of dislocatory heavings and collisions, terminating in deathlike silence and the overthrow of the car which we occupied, gave certain evidence that we had gone off the line.[45]

Not only back East, but out West as well. Charles Weld again:

The locomotive is very unlike ours, being an uncouth looking machine with a prodigious bottle-nose chimney and an iron-barred vizor-like affair in front, called a cow-catcher, though, as I can attest from observation, it is not at all particular as to the kind of animal it catches.

As the railways, with few exceptions, are unprovided with fences, the herds and flocks are at liberty to roam on the track; sheep, especially, are fond of resorting to the line at night, which they find drier than the damp clearings. These animals, however, are not deemed formidable obstacles. An engine cleverly dashed through a flock of 180, the greater portion of which were summarily converted into mutton.[45]

American trains were stopped by brakemen applying individual brakes in each separate railroad car. If they failed to act simultaneously, or there was a particularly heavy load—disaster. In Massachusetts in 1848, for example, there were fifty-six deaths and sixty-five serious injuries—more than one major accident per mile of railroad track. Railroad accidents came to be regarded as a natural hazard much as automobile accidents are regarded today.

The man who solved the accident problem was George Westinghouse who, in 1869, took out the first of 103 patents relating to the air brake. Westinghouse had been in an accident himself and realised what was needed: a system for applying all brakes to all wheels at the same time. After several early attempts to provide the solution, he learnt how the engineers of the Mont Cenis Tunnel in Switzerland were using drills operated by compressed air, supplied through a length of pipe. This gave him his idea.

An air-compressor fitted to the locomotive and operated by a small steam engine could supply compressed air to a reservoir.

George Westinghouse, 1846–1914

From the reservoir, compressed air could be carried under the length of the train by a flexible pipe, to brake cylinders on each coach. On the release of air, a piston rod fitted to each cylinder would be pushed against the brake shoes to push against the wheel and stop the train. The entire system was controlled by the driver, who released the flow of compressed air from the reservoir merely by operating a valve.

For all its brilliance, this system did not solve the problem of how to stop individual coaches if they broke away from the rest of the train, or if part of the system failed and interfered with safe operation of the brakes elsewhere. To solve this difficulty, Westinghouse put the system into reverse. Instead of using the compressed air to work the brakes, he used it *not* to work them. The compressed air was now employed to keep the brakes *away* from the wheels but, as soon as the supply of air was cut off, through some fault or accident, the brakes would automatically fall against the wheels and stop the train or, if it had broken loose, an individual coach.

To prove his point, and sell his invention, Westinghouse made comparative tests throughout the country with a locomotive carrying fifty freight wagons. At 20 mph, the hand-brake system brought the train to a halt in 794 feet. The 'Atmosphere Air Brake' did the same thing in 166 feet.[52] Within three years of its conception, the invention was in general use. Railroad accidents through inadequate brakes were, almost, a thing of the past.

The same year that Westinghouse invented the air brake, the transcontinental railroad was completed. On 10 May 1869, at Promontary Point, Utah, the last spike was driven into the ties of the Union Pacific—a spike made of California gold. The railroad had been completed at terrific speed and terrific human cost. Three thousand Irish, ten thousand Chinese and an assortment of ex-soldiers and ex-convicts had fought off attacks by Indians and pushed the line forward over mountains, across deserts, in blistering heat or sub-zero temperatures, at a rate of eight miles a day for five years. The advance was eagerly reported by the new telegraph authorised to follow the same route. When the final news came through, the Independence Bell was rung in Liberty Hall and Chicago held a parade that was seven miles long. Throughout the country there were flags, bands and singing as America celebrated her ambitions and achievements in popular song. As she had done before and has done since—more than any other nation. A mood indicator. And the songs themselves along with the feelings and events they celebrate pass into the national consciousness.

Work crew, Union Pacific Railroad

Above and opposite, two views of Promontory Point near Ogden, Utah, just before noon, 10 May 1869—'Officials, guests and construction forces await the ceremony that will link the nation.' On the right the Union Pacific's engine No. 119 and, on the left, the Central Pacific's 'Jupiter' No. 60. The gold and silver spikes were driven by Leland Stanford and Thomas C. Durant.

Union Pacific's 'Pony' locomotive No. 3

Far right: Dale Creek Bridge crossed by the first Union Pacific engine, 1868

Songs about the Telegraph ('Hello My Baby: send me a Kiss by Wire'), and songs about spectacular feats of engineering ('Strolling on the Brooklyn Bridge'). Not only the celebration of achievements but the romantic image of faraway towns, of getting to new places—whether it's 'Chicago, Chicago' or 'Eighteen Miles to Tutumcari'. The physical and mental restlessness of Americans. In all ages. The songs of the early pioneers updated into the novels and poems of the 1950s' Beats—Kerouac and 'On the Road'—or the motorbike movies and car cinema of the late 1960s—'Easy Rider', 'Two-Lane Blacktop'. Combining the achievement *and* the romance, the songs of the railroad:

> Singing through the mountains
> Buzzing o'er the vale
> Bless me, this is pleasure,
> A-riding on a rail.

Songs about Casey Jones, construction hold-ups, engineers, particular lines—'The Aitcheson, Topeka and the Santa Fe', the 'Chatanooga Choo Choo'. The numbers are legion. Both genuine folk and more recent pop. And accidents:

> A mighty crash of timber,
> The sound of hissing steam,
> The groans and cries of anguish,
> A woman's stifled scream.
> The dead and dying mingled
> With broken beams and bars,
> An awful human carnage—
> A dreadful wreck of cars.*

No one in Britain sang about getting to Wolverhampton, Aberdeen or Potters Bar, and not very many about the London, Midland and Scottish Railway. Or the little that was sung has been long forgotten because the established nations of Europe were denied that tremendously invigorating feeling of belonging to a nation that was opening up and on the move. For America, the tradition and some of the feeling is still there—'Up, Up and Away' or 'I'd like to get you on a Rocket ship'—though the emotions are getting a bit jaded.

In 1869, the Union Pacific was a symbol.

There will speedily be other railroads across our continent. The rivalries of sections, the temptations of commerce, the necessities of our political system, will add at least two more through lines within a

*The Chatsworth wreck occurred in Illinois in 1887. Eighty-two people died.[53]

generation's time. But this, the first, will forever remain the one of history; the one of romance. Its construction in so short a time was the greatest triumph of modern civilization.[54]

Its completion was more than just another railroad finished. It was the new Lewis and Clark; it was 'Manifest Destiny' fulfilled. The back of a great continent had been broken.

In 1870, the first transcontinental train began its pioneer journey from Boston to San Francisco. This time pioneer discomfort was not a part of the experience—at least for the directors of the Union and Central Pacific Railroads, who rode a coach provided free by publicity-conscious George Mortimer Pullman (1831–97).

Pullman's publicity, like Singer's advertising, had its reward. The delighted directors decided 'that there will be no delay in placing the elegant and homelike carriages upon the principal routes in the New England States'.[31] Pullman was a mid-century man—possessing drive, imagination and determination, and on the lookout for any way to turn a fast buck. Like other mid-century men—Morse, Singer, the big bosses of the railroads, even McCormick—he made money less from his own inventions than from the clever organisation and development of other people's.

Sleeping cars were already running on the railroad between Philadelphia and Baltimore in the 1830s—'Complete in every convenience', though not in the opinion of the travellers who found them cramped and uncomfortable. T. T. Woodruff patented a sleeping car with a three-tier berth in 1856, but the break through the comfort barrier began with Pullman's converted railroad cars in 1858. Pullman was an engineer. At the age of twenty-four he had had a contract from the city of Chicago to raise a number of streets above flood level. With the profit, he remodelled two railroad cars bought from the Chicago and Alton Railroad company and produced his first car—carpetless, sheetless and lit by candles. The project failed.

Pullman opened a general store in booming Colorado, made $20,000 and, after four years, was back in Chicago ready to try again. The result was 'The Pioneer'—fortunately more original in its features than in its name. Instead of bunks suspended from the ceiling, the upper berth was formed by beds that folded out from the wall. The idea was Woodruff's, but Pullman used only two tiers instead of three. The lower berth was still converted from the daytime seats but, instead of the backs being lowered, two seats were slid together. And the cars rode on springs additionally cushioned with rubber. Evans, half a century earlier, had predicted accurately, 'the passengers will sleep in these stages

George Mortimer Pullman (1831–1899) and his 'Pioneer' sleeping car

The original Pullman sleeping car with the upper berths already lowered for use. The occupants are posed to show how much head room there is in the lower berth.

as comfortably as they now do in stage boats'.[43] The Pioneer's gimmick was 'luxury': lavish, even exotic decoration, curtains, mirrors, chandeliers, black walnut panelling with marquetry in-lay—and space. The cars were fifty-four feet long, a foot wider and two and a half feet taller than any previous car. Cost: $20,000.

My contribution was to build a car from the point of view of passenger comfort; existing practice and standards were secondary.[1]

This cavalier rejection of existing standards meant that 'The Pioneer' had clearance problems: it couldn't get under bridges,

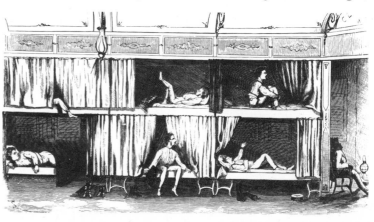

Typical two-tier sleeping car—Jack Lemmon and Tony Curtis are in the upper berths on the right. Marilyn Monroe is in the closed berth on the left.

down tunnels or past station platforms. It was immediately dubbed 'Pullman's Folly', till fate, in the form of an assassin's bullet, gave Pullman his break. The nation, in paying tribute to the dead Abraham Lincoln—and under pressure from Mrs Lincoln—requested the use of the biggest and best railroad car in the country to transport his body from Chicago to Springfield. The necessary engineering work on the line, to accommodate the huge car, was quickly done and 'The Pioneer' became famous overnight. Two years later, there were forty-eight cars operating on three railroads, providing comfort, profit and an unexpected impetus for the standardisation of railroad gauges.

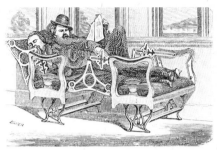

Blood's Patent Railroad Car Seats (1871)
– for those who couldn't get a Pullman

Pullman followed the success of the sleeping car with the hotel car and the dining car. The 'Delmonico', Pullman's first diner, was put in service in 1868. Its kitchen was sited in the car's middle so that, in Pullman's words, 'in whatever direction the car may be travelling, at least half of it will be in advance of the kitchen, the odours of which are borne by the draught toward the rear'.[58]

Until the advent of the diner, railroad eating had been a problem. The usual way was for the railroad to make a twenty-minute stop:

. . . all the doors are thrown open and out rush all the passengers like boys out of school, and crowd round the tables to solace themselves with pies, patties, cakes, hard-boiled eggs, hams, custards and a variety of railroad luxuries too numerous to mention. The bell rings for departure, in they all hurry with their hands and mouths full and off they go again.[44]

What Captain Maryatt didn't know was that the eating-house owners were usually in cahoots with the train crews so that passengers, as soon as they'd paid the standard charge of four bits and begun eating, were called back to the train so that the food could be saved for the next customers. The practice, honest or crooked, was neither good for the digestion nor the temper and set a pattern for stand-up, hurried eating that has proliferated in the era of the drug-store counter and the automat. In contrast, Pullman's Dining Cars made possible leisured eating of European cuisine in peace and comfort.

The choice is by no means small. Five different kinds of bread, four sorts of cold meat, six hot dishes, to say nothing of eggs cooked in seven different ways and all the seasonable vegetables and fruits, form a variety from which the most dainty eater might easily find something to tickle his palate and the ravenous to satisfy his appetite. . . . To breakfast, dine and sup in this style while the train is speeding at the

Dining Car Menu on the Baltimore and Ohio Railroad—the wines were imported

rate of nearly thirty miles an hour is a sensation of which the novelty is not greater than the comfort.[59]

Pullman, like Singer, created a market almost before the market realised it was there. Luxury was something new, but it very soon became identified with both aspiration and achievement, national and individual. With the Pullman car, America had arrived; the American was no longer the unsophisticated backwoods hick. And, unlike in Europe, luxury had been achieved within the context of democracy. The French had created a luxury train in 1857—exclusive to Napoleon III; in 1865, Pullman was offering the same luxury for everyone. $2 a night: just half a dollar more than in the ordinary sleeping cars. On the other hand, Pullman's innovation was, in one sense, divisive.

No royal personage can be more comfortably housed than the occupant of a Pullman car, provided the car be a Hotel one.[59]

The significant word is 'provided'. America had been very careful to preserve equality on the railroad by not introducing the European system of dividing accommodation into first, second and third class. 'The rich and the poor, the educated and the ignorant, the polite and the vulgar, all herded together.'[11] Nevertheless:

Though in theory all are equal there are practically various classes of passengers. On the main lines there are cheap trains for immigrants. There are attached to most trains 'drawing-room cars', 'reclining-chair cars', and 'sleeping cars' or night cars for which additional charge is made. Some of these select cars belong to speculating builders or companies who purchase the privilege of attaching them to the trains and make their profit by the additional charge. The most notable of these speculations are the Pullman 'Palace Hotel' cars.[56]

And, for the rich American:

An additional zest is given to the good things by the thought that the passengers in the other cars must rush out when the refreshment station is reached and hastily swallow an ill-cooked meal.[59]

Everyone could be royalty—if he could afford to pay.

The Pullman cars were mass-produced. And, for the first time, mass-production could mean mass luxury. The American System of Manufacture responded successfully to the new demand and showed that it could not only produce fast and cheap but could put a really high surface gloss on the basic technology. That it could achieve by machinery what could previously only have been done well by hand. The business and manufacturing implications were enormous. New markets beckoned: the US could now compete even in the craft markets that were traditionally the preserve of Europe. By 1880, in a true coals-to-Newcastle reversal of roles, Pullman luxury was running on the smooth rails of Britain.

The last great area of the American West to be reached by railroad was the Southwest: cattle and cowboy country—and, for a short while longer, Indian and buffalo country. General Sheridan's pronouncement sounded the death-knell:

Let them kill, skin and sell until the buffalo is exterminated as it is the only way to bring lasting peace and allow civilisation to advance.[32]

Between 1872 and 1874 four million buffalo were killed by the white hunters of Manifest Destiny. The buffalo was the traditional Indian livelihood. No buffalo—no Indians. But 'civilisation' continued its advance.

The process had begun with the Mexican War in 1847, which was won for the US with the help of Colt's revolver. After the war Spanish cattle, along with the buffalo, ran wild and bred on the open range. It was free grazing for the cattle, and a livelihood for the cattle ranger. But as the settlers moved farther and farther west, lured by the government promise of free land, they brought their farms on to the plains. There was a clash of interests. The farmer was taking the ranger's country, the rangers' cattle were trampling the farmers' wheat. There were fights—especially in areas such as Oklahoma, where the two ways of life overlapped. What the farmer desperately needed in a country where there were few rocks and no trees for fencing was something 'horse-high, bull strong and pig tight'.[31] And cheap fencing, according to a government report of 1871, cost up to three hundred per cent more in the West than in the rest of the US.[38] Ordinary wire proved useless till, in 1874, Joseph Glidden of Illinois came up with the answer:

Wire cutters on Brighton Ranch, Custer County, Nebraska, 1885. According to the contemporary caption, the settlers, taking the law into their own hands, cut 15 miles of wire and posed for the photograph afterwards.

A twisted fence wire having the transverse spur wire D bent at its middle portion about one of the strands A of said wire fence and clamped in its position and place by the other wire Z, twisted upon its fellows substantially as specified.[38]

It was widely advertised:

Safety to passengers and property. Lasts twice as long as any other kind of fence. Sparks do not set it alight. Floods do not sweep it away. Its merits commend it as the best fence in the world.[16]

Glidden wasn't, of course, the first inventor of barbed wire—as with so many inventions, there had been other work to anticipate his—but he was the man who made barbed wire commercially viable and who had to fight the bulk of the inevitable patent battles. One hundred and fifty railroad companies used it for keeping cattle off the track. Farmers, initially sceptical, used it for keeping cattle out of the wheatfields. The harvest rose to become one-sixth of the world's grain output. The people who lost out were the cowboys. Their resistance made banner headlines in the local papers: 'Wire cutters cut 500 miles in Coleman Country'.[38] But it was a futile battle.

They say that Heaven is a free range land,
Good-bye, good-bye O fare you well;
But it's barbed wire for the devil's hat band;
And barbed wire blankets down in hell.[38]

If barbed wire hastened the end of the open range, the decline had already begun. In the 1850s and 1860s, the comparatively few cattle that went east were driven along hundreds of miles of trails to the slaughterhouses of the new Midwestern cities like Cincinatti, 'the pork shop of the Union'.[44] But in 1867 the entire business was changed by one man: J. G. McCoy, the original 'real McCoy', an Illinois livestock shipper born in Springfield in 1837. McCoy was shrewd enough to see that the burgeoning cities of the east could do with cheap meat and that the cattle rangers of the Southwest could supply it. If, somehow, he could bring Texas and the East Coast cities together in Illinois he would do a service to all three communities—and to himself.

The longer the idea of this enterprise was harboured, the more determined he became and the more enthusiastic to carry it out. It became an inspiration almost irresistible, rising to all other aspirations of his life.[62]

McCoy, referring to himself here in the third person, told his own story with the campaigning fervour usually reserved for the religious. And the idea, when it came, was like a salvation to him: the railroad.

The plan was to establish at some accessible point a depot or market to which a Texan drover could bring his stock unmolested, and there, failing to find a buyer, he could go upon the public highways to any market in the country he wished. In short, it was to establish a market whereat the Southern drover and the Northern buyer would meet upon an equal footing, and both be undisturbed by mobs or swindling thieves.[62]

The old cattle trails and the new railroads crossed at Abilene, Kansas:

A very small, dead place consisting of about one dozen log huts . . . and, of course, the inevitable saloon . . . Abilene was selected because the country was entirely unsettled, well watered, excellent grass and nearly the entire area was adapted to holding cattle.

McCoy built his stockyard in eight weeks.

When it is remembered that this was accomplished in so short a time, notwithstanding the fact that every particle of material had to be brought from the East and that, too, over a slow-moving railroad, it will be seen that energy and a determined will were at work.

Round-up of cattle—'Cutting out a steer', a wood engraving by Frederic Remington. Remington (1861–1909), journalist and illustrator, was one of the first East-coasters to popularise the cowboy.

McCoy had a stockyard, but no stock.

A man well versed in the geography of the country . . . was sent into southern Kansas and Indian Territory (Oklahoma) with instructions to hunt up every straggling drover possible (and every drover was straggling for they had nowhere to go) and tell them of Abilene.

Good news for the drover, or was it?

It was almost too good to be believed. Could it be possible that someone was about to afford a Texan drover any other reception than outrage and bribery? They were very suspicious that some trap was set.

For good reasons: rustling was the drover's occupational hazard.

If the mob could not frighten the drover until he would abandon his stock or if they failed to obtain a pretext for killing him outright, resort was had to stampeding the cattle . . . the entire herd would be greatly injured and many of the cattle utterly ruined. Whilst the drover was engaged in regathering the cattle, the members of the mob would be as busy secreting all they could find.

The scenario for a thousand Hollywood Westerns.

In the first year of McCoy's operation, the drovers risked their necks in the drive to Abilene to the tune of 35,000 cattle. From that moment, the southwest meant meat just as the Midwest meant grain. And McCoy, an American hero as much as Singer or Whitney or Lowell, was also one of the first men to give to the world America's most famous legend—the legend of the cowboy. He wrote in 1873:

Loading cattle at McCoy's Stockyards, Abilene, Kansas, 1867

The cowboy lives hard, works hard . . . has but little if any taste for reading. He enjoys a coarse practical joke or a smutty story; loves danger but abhors labour of the common kind. He would rather fight with pistols than pray; loves tobacco, liquor, and women better than any other trinity. His life borders nearly upon that of the Indian.[62]

A character sketch, suitably cleaned up and romanticised, for the same Hollywood scenario.

To supply the new demand, the cattlemen began to stock the range. Meat meant money. New animals were introduced to replace the native Longhorns; the newfangled barbed wire made excellent breeding compounds. By 1876, the southwest was pegged down by lines of metal: barbed wire, telegraph wires and railroad tracks. The Indians were pegged to reservations. The drovers lost their jobs. The public gained tender beef; instead of building up muscle on the long walks, the cattle relaxed on the train. Abilene declined as the railroad pushed west to 'Shootin'' Newton and Dodge City, but Chicago boomed—'Boss' town of America, 'Foodsville', USA. Henry Fuller (1857–1929):

The only great city in the world in which all its citizens have come for the avowed object of making money.

From a one-eyed cow town of 1827, supplying food to outlying military garrisons, it had grown to become the centre of a network of nine railroads. John Wright, in 1868:

She is, in truth, the artificial hub of the North West and, as such, of the   *Union stockyards, Chicago, 1866*

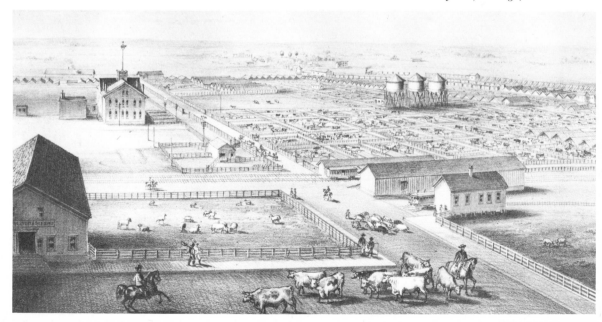

79

Republic. Her railway spokes fasten her fellows to her so securely, that no rivers can wash them away; the wheel revolving with resistless power so that no interposing wheel can come into resistance.

There were 6980 miles of railroad track in Illinois, 1000 more than in any other state in the Union, and even one hundred years later, after the railroads' decline, Chicago's O'Hare Airport remains the busiest in the world.

By mid-century, Chicago was the manufacturing headquarters of McCormick reapers, Pullman cars and other industries attracted by the growing western market. Cattle were now to confirm it as the undisputed centre of the nation's biggest industry—the food industry.

The rise of the cattle market was phenomenal. In 1867, 35,000 longhorns were delivered to McCoy's Chicago stockyards. In 1868, 75,000; 1869, 350,000. By 1880, almost four and a half million cattle had reached Chicago in ten years. But none of the new arrivals went to McCoy. He, like so many American innovators, had gone bankrupt.

Most cattle were for onward shipment, by rail, to East Coast butchers, but soon the more business-minded saw profit in slaughtering on the spot. Or, as beef baron Philip Armour later put it, '. . . to turn the bristles, blood, bones and the insides and outsides of bullocks into revenue'.[31] Mass-production techniques, now familiar throughout most of American industry, were introduced into the slaughtering business in the form of a primitive assembly line, or rather—as cows, sheep and pigs were being dismembered—a dis-assembly line. There was little mechanisation, but there was a 'factory' system. The animal was caught, roped, pulled to the ground and speared or shot. It was bled, part of the hide removed and the carcass hoisted by a hand windlass to be washed, cut up and pushed into the cooler by men on portable ladders. The aim of the business was to ship meat already dressed, quickly and cheaply, in order to undercut the established East Coast butchers. The problem was that shipment was limited to the winter months. Shipped at any other time, the meat went off. This situation was rescued by the 'reefer'—the Refrigerator car.

Natural ice was used in connection with summer slaughtering in 1857, but the first refrigerator car patent was issued ten years later. Others swiftly followed. By 1872, meat packer George Hammond was using refrigerator cars for shipments to Boston, but the ice melted, and at twenty-four-hour intervals the car had to stop in a siding to reload with fresh ice. The ice also discoloured the meat. Soon butcher shops were displaying signs reading 'No Chicago Beef Here'.

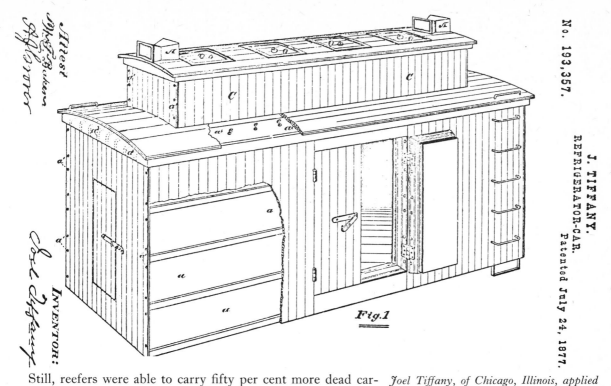

Attest

Inventor:

Fig.1

Still, reefers were able to carry fifty per cent more dead carcasses than ordinary cars could carry live ones. Hammond's
economic success was sufficient to bring East Coast butchers like
Swift and Armour to the Midwest, and it was Swift, and his
engineer Chase, who produced the first really reliable car. Ice was
piled on the car roof to produce cooled air which became heavy,
dropped to the floor and was expelled as warm air through ventila-
tors. Meat thus cooled was fresher on arrival. The Chicago meat
business was freed of seasonal limitations, and slaughtering
became mechanised. Pens with tilt-up floors for killing and moving carcasses; overhead rails for hanging and transporting to the
waiting reefers. And the Chicago slaughterhouses, like the Lowell
factory before them, became one of the main sights for visiting
Europeans—'*The* wonder of the New World'. In 1898, the
phenomenon was witnessed by Rudyard Kipling:

I climbed to the beginning of things and, perched upon a narrow beam,
overlooked very nearly all the pigs ever bred in Wisconsin. They had
just been shot out of the mouth of the viaduct and huddled together in
a large pen. Thence they were flicked persuasively, a few at a time, into
a smaller chamber, and there a man fixed tackle on their hinder legs so
that they rose in the air suspended from the railway of death. Oh! it
was then they shrieked and called on their mothers and made promises
of amendment till the tackle-man punted them in their backs, and they
slid head down into a brick-floored passage, very like a big kitchen sink
that was blood red. There awaited them a red man with a knife which
he passed jauntily through their throats, and the full-voiced shriek became a sputter, and then a fall as of heavy tropical rain. The red man
who was backed against the passage wall stood clear of the wildly

*Joel Tiffany, of Chicago, Illinois, applied
for a patent on his Refrigeration Car in
1876. His car was one of the first that
actually worked.*

81

kicking hoofs and passed his hands over his eyes, not from any feeling of compassion, but because the spurted blood was in his eyes, and he had barely time to stick the next arrival. Then the first stuck swine dropped, still kicking, into a great vat of boiling water, and spoke no more words, but wallowed in obedience to some unseen machinery, and presently came forth at the lower end of the vat and was heaved on the blades of a blunt paddle-wheel-thing which said, "Hough! Hough! Hough!" and skelped all the hair off him except what little a couple of men with knives could remove. Then he was again hitched by the heels to that said railway and passed down the line of the twelve men—each man with a knife—leaving with each man a certain amount of his individuality which was taken away in a wheelbarrow, and when he reached the last man he was very beautiful to behold, but immensely unstuffed and limp . . .[63]

The introduction of machines for de-hairing, cutting, grinding and cooking were the essential prerequisite for the allied growth industry: canning.

'The Art of Preserving Animal and Vegetable Substances for Several Years' was an established European practice. It had been used as a method of stockpiling food for Napoleon's armies in 1810, and was used by the British Navy in 1818 for preserving fish and soup. It consisted of sealing food in bottles which were then submerged, cooked and made airtight in boiling water. Occasionally, in place of bottles, tin cans. British Arctic explorers had preserved food in cans as early as 1815, and in America in 1819, Thomas Kensett received a patent for 'an improvement in the art of preserving by using vessels of tin'. It was Kensett who described his process as 'canning'. But the most important American contribution was cooking and canning along mass-production lines, stimulated by the need to feed thousands of soldiers during the Civil War. Normal processing time was six hours. In 1861, Isaac Solomon, a Baltimore canner, introduced calcium chloride to the water to increase the temperature to 240 degrees Fahrenheit, thereby reducing cooking time to twenty-five minutes. This, coupled with improved know-how in manufacturing the cans themselves, resulted in output leaping from 2000 cans a day to 20,000. By 1872, cans had virtually ousted bottles, and a small firm called Libby, McNeill and Libby used them for packing a newfangled food—corned beef.

The American diet, as a result of the railroad, the reefer and the can, was transformed. In the 1840s, Robert Stevens's Camden and Amboy line was already nicknamed the 'pea' line as it carried peas, strawberries and other fruit and vegetables from New Jersey, 'The Garden State', to New York and beyond. In the winter

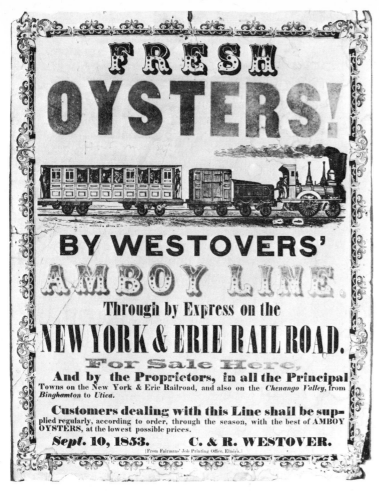

of 1852, peas and tomatoes for a shivering Chicago were being dispatched from New Orleans. The Midwest was getting lobsters and oysters from New England by 1865. By 1870, thanks to the refrigerator car, perishables like peaches from Georgia, bananas, grapes, pineapples, oranges and lemons from California, travelled across the breadth of America. The city of New York, one hundred years after Independence, was receiving food fresh, frozen and canned, from forty-two states. At the new low prices, even the poor could afford meat twice a week.

All of this has been done in the early morning of the country's life. . . . The day lies before it as the future. What will not this people accomplish during their day.

Chapter 3

109th US Colored Infantry, Fort Corcoran, Arlington, Virginia. The South rejected all proposals to use slaves as soldiers in case it opened the door to freedom. Congress in the North, so sensitive to the feelings of the border states, also refused to enlist negroes till 1862 when fugitive slaves were regarded as 'contraband of war' and put to work.

Civil War

Eighty years after Independence, America went to war with itself. On 4 February 1861, at the Congress of Montgomery, Alabama, Georgia, South Carolina, Florida, Mississippi and Louisiana decided to elect Jefferson Davis President of the Confederate States. In the months that followed, five more southern states joined the Confederation. President Lincoln, in the North, fought to preserve the Union.

The clash between the North and the South was inevitable. It was the clash of two rival social and economic systems that were the direct products of different technologies. The sleepy South versus the dynamic North— the cotton gin versus the reaper. Man power versus machine power.

The South, happily dependent on cotton, with a ready supply of free black labour, moved slowly westward. The rotation of crops to preserve the land was unnecessary; land, after all, was plentiful. The movement left behind it a trail of ruined soil. The North, rich on the products of its mechanised industries, looking for new markets, resented the South's waste of natural resources and its refusal to mechanise its own industries. So, too, did a few Southerners:

Unquestionably, wealth develops wealth, energy develops energy, talent develops talent. What then, must be the conditions of those countries which do not possess the means or faculties of centralizing their material forces, their energies and their talents? Are they not destined to occupy an inferior rank among the nations of the Earth? Let the South answer?[64]

The two systems collided in the west: 85

Tennessee, Kansas, Nebraska. Ostensibly an argument over the issue of slave-owning and slave-free states, the conflict became a debate over the meaning of liberty within the terms of the Constitution, and the right of each state to determine its life-style for itself.

The technological as well as physical isolation of the South had steadily increased since 1820. Railways ran east-west, not north-south. Almost all technological innovation had come from the North, and even the bombastic popular press recognised that technology would determine the war's outcome.

Take our word for it, these Yankee geniuses will yet produce some patent secession-excavator, some traitor-annihilator, some rebel-thresher, some Confederate state milling machine, which will grind through, shell out or slice up this war, as if it were a bushel of wheat, or an ear of corn or a big apple. [*Philadelphia Enquirer*][2]

The American Civil War showed the ease with which the technology of peace—the telegraph, the railways, the sewing machine which mass-produced uniforms—could be converted into the technology of war. It showed how war itself could provide the catalyst for invention—submarines, torpedoes, battleships—and guns. Since Independence, the development of American armaments had progressed unsteadily.

Under these circumstances, I cannot forbear to reiterate the recommendations which have been formerly made, and to exhort you to adopt with promptitude, decision and unanimity, such measures as the ample resources of this country can afford, for the protection of our seafaring and commercial citizens.

President John Adams addressed the US House of Representatives on 19 March 1798. The result was guns, the 'Whitney System' and self-sufficiency.

Perhaps no country ever experienced more sudden and remarkable advantages . . . than we have derived from the arming of our marine protection and defense.

After another sharp reminder of the need for self-sufficiency in the war with Britain in 1812, the United States determined again to increase her ballistic know-how. She still did not manufacture her own cannon. After 1819, she did. The first ones were produced at the West Point Foundry by the man who made the boilers for the experimental railroad engine 'Tom Thumb'. By 1860, on the eve of the Civil War, there were three foundries in the North, soon seven. But still West Point was the most important and its most important product was the Dahlgren gun.

Invented by a naval lieutenant and made from cast iron, its importance lay in its shape. Dahlgren estimated that both greater fire-power and greater safety could be achieved if the thickness of the gun's walls was not uniform: thickness at any one point should be determined by the amount of pressure that part of the iron walls would have to stand. As this was greater at the breech end where the explosion occurred, and slight at the muzzle end, Dahlgren produced his revolutionary-shaped cannon 'like a soda water bottle'.[65]

Because of its smoothness and simplicity, the gun could be cast in such a way that the danger of fractures from the internal explosion could be greatly reduced. Also, because the wall strength was now where it was most needed, a greater charge could be used and solid iron cannon balls could be fired farther than ever before. President Lincoln was given a demonstration.

The splendid course of the 11 inch shell flying through 1300 yards of air, the lighting, the quick rebound and flight through the target . . . were scenes as novel and pleasant to me as to all the rest of the party. The President was delighted.[66]

But so fast is the movement of war that, by 1863, the *Scientific American* proclaimed 'the day for smooth-bored cannon has gone by, all artillery must henceforth be rifled'. The Dahlgren was out of date.

The North, in need of even more penetrating fire-power, began to replace solid shells with exploding ones. The trouble was that the shells exploded the guns as often as they exploded their targets. The solution was provided by Robert P. Parrot. Based on an idea that originated with Krupp in Germany, the Parrot gun had a rifled bore but was more conventionally gun-shaped; its great innovation was to have huge hoops of wrought iron shrunk on to the breech for reinforcement. It was immediately brought into service.

15-inch Dahlgren muzzle-loader

White Point Battery, Charleston, 4 December 1863—two days before the start of the Christmas bombardment. On the fortress walls Dahlgren muzzle-loaders, on the left, a version of the Parrott. In the distance, right, on an island guarding the entrance to Charleston harbour, is the hazy outline of Fort Sumter. After South Carolina had seceded from the Union in 1860, Fort Sumter became a US (Federal) outpost. On 12 April 1861 Confederate guns, watched by 'the beauty and fashion' of Charleston, fired the opening shots of the Civil War. Fort Sumter surrendered the following day. The picture was painted in oils by C. W. Chapman.

The 'Swamp Angel', an 8-inch, 200-pound Parrot gun, opened fire on Charleston on the night of 22 August 1863. Alarm and confusion amongst the citizens, but of the thirty-six shells fired six exploded in the gun itself, and the last blew out the breech. Obviously there was room for improvement. Four months later, on 6 December:

We are daily tossing twenty shells or so into the city of Charleston.[67]

And on 29 December:

Charleston had a very merry Christmas. One hundred and thirty shells had been thrown into her up to the afternoon church time.[67]

The new gun in question was a 30-pound Parrot mounted on a wooden carriage on Cummings Point. The effectiveness of the iron hoops was being proved.

Sixty-nine days elapsed between the first and last discharges of the gun. It was being fired the four thousand six hundred and sixth round when it burst.[1]

There was also a revolution in rifle design. Guns had been breech-loaded for centuries, but during the eighteenth century the improvement in gunpowder made firing of breech-loaders extremely dangerous. The design had lapsed in favour of the muzzle-loaders like the Kentucky. In 1849, the design was re-

Union breech-loading Parrot, Atlanta, Georgia, 18 July 1864

88

Gatling and his machine gun

vived by Chambers' 'Breech-Loading Ordnance' in which the plug in the breech was fixed by a threaded screw. In 1857, the first practical breech-loader was patented by Sharpe: the 1859 'Sharpe-shooter' could be fired eight times a minute. In 1862:

So rapidly did our boys fire into the rebels, they were completely discomforted and forced to retire.[65]

The speed of the Sharpe was enough to transform land warfare.

But the most revolutionary of all American guns was refused official approval until the war was over.

UNITED STATES PATENT OFFICE
Specification forming part of letters patent No. 36,836, dated November 4. 1862.
To all whom it may concern:
Be it known that I, RICHARD J. GATLING, of Indianapolis, county of Marion, and State of Indiana, have invented new and useful improvements in Fire-Arms, and I do hereby declare that the following is a full and exact description thereof . . . the object of this invention is to obtain a simple, compact, durable and efficient firearm for war purposes.

There had been versions of the machine gun a few years before. In 1861, J. D. Mills was advertising ' "The Union Repeating Gun"—an Army in six feet square'. In June 1862, Captain Williams of the Confederate South produced a gun that fired an unprecedented forty rounds a minute but failed to stampede the enemy, who were intrigued rather than frightened. Gatling's gun, basically similar, was an enormous improvement.

The design was simple. Round a crank-operated central shaft were fixed ten or more breech-loading rifles. As the shaft was rotated by one man, another loaded each individual gun with cartridges. In 1862, the machine gun could fire 350 rounds a minute. Gatling wrote to President Lincoln in 1864, assuring him of the brilliance of his invention. In 1866, the US Army finally bought a hundred—a year too late for the war. But by 1882 the machine gun could fire 1200 rounds and was ready for the next one. It proved itself very effective against the Indians, and went on to score a triumph in World War I.

Gatling was a doctor. He justified his invention on medical grounds. The carnage was greater, but cleaner, and in the long run saved more lives by ensuring the war would end more quickly. It was the same justification that the allies were to use when they dropped the atom bomb on Japan.

The South's first reply to the North's superior fire-power was the *Merrimac*—the iron ship. It was based on a design made some forty years earlier by John Stevens, the father of T-rail Robert. Stevens had built his own steam locomotive in 1820; his ship *The Phoenix* had been the first ocean-going steamboat. But his design for a floating fort had been rejected. The design was later improved by a proposal that the entire fort be encased in iron. In the 1840s the design was again rejected as too extreme. It was resurrected by the Confederates, because it was only by the use of extreme ideas that they had a chance of victory.

The *Merrimac* was a recycled US warship—real name, *The Virginia*. In place of her original wooden superstructure, the Confederates erected what amounted to a 7-inch-thick wooden gun emplacement covered by two layers of $2\frac{1}{2}$-inch-thick iron plates, angled to 35 degrees. She was steam driven, weighed 3500 tons, and armed with a cast-iron ramming prow and 9-inch Dahlgren cannon, she wreaked havoc—actual and emotional.

We have only two guns that can make an impression on her . . . if she is invulnerable, there is no reason why she should not steam up the Narrows and lay the city [New York] under contribution.[67]

The warships of the Union had no answer to her. The *Merrimac*'s guns smashed into their wooden walls whilst the cannon balls from the guns turned against her made a few dents in the iron sheeting and fell harmless into the sea. For Rear Admiral (once Lieutenant) Dahlgren, now Chief of the US Navy Yard,

It was a serious business and if the Merrimac were successful no one could anticipate the consequences to our side.[65]

The 'Merrimac' sinking the 'Cumberland', 8 March 1862

The sinking of the *Cumberland* on the Potomac River led President Lincoln to believe a Confederate shelling of the White House not only possible but imminent. The South ordered the construction of four more ironclads. The North searched desperately for a solution. It was this need for a missile that could penetrate iron that hastened the advent of the exploding shell and the Parrot gun.

But the immediate answer to the *Merrimac* was the North's 'cheesebox on a raft', the *Monitor*. It was the brain-child of John Ericsson (1803–89), a 59-year-old Swede who, before arriving in the US in 1839, had competed against Stephenson's 'Rocket' in the locomotive trials on the Manchester-Liverpool Railway. His revolutionary ship was another version of the floating fort but, in almost every respect, a more radical design. A shallow iron hull was covered by a flat, overhanging iron sheet—the effect was rather like a large, upturned saucer over a small dish. On top of the saucer was a box-like iron gun emplacement which was to contain two Parrot guns. She was driven by a screw propeller, like her rival, and was so low in the water that her decks were permanently awash. She was built in four months, launched on 15 February 1862, and immediately put to sea. On 9 March she

The crew of the 'Monitor' grouped round the gun turret and the 'Monitor's' revolving turret showing the emplacements for the Dahlgren guns, James River, 9 July 1862

The great fight between the 'Merrimac' and 'Monitor', 9 March 1862. The 'Monitor' with its revolving gun turret is able to use both its guns, whereas the 'Merrimac' is restricted to one.

encountered the much larger and more heavily armed *Merrimac*. It was six Dahlgrens against two—the *Monitor*'s Parrots weren't yet ready. But the *Monitor* had one unique feature: her gun turret revolved. That meant that whatever the position of the *Monitor* and whatever the manoeuvres of the *Merrimac*, the eleven-inch guns would always be on target. Two guns could now do the work of an entire battery.

The Monitor was, without doubt, the most remarkable production of the constructive art that appeared during the war. . . . Too much credit cannot be awarded to Captain Ericsson for his brilliant conception of this floating battery and the navy must be ever grateful to him for preserving it from dire disaster which was averted by [its] appearance at the moment of crisis.[69]

The battle of Hampton Roads, fought on 8 and 9 March 1862, was indecisive. Neither vessel did permanent damage to the other, but the battle marked the end of an epoch and the beginning of another. It was the end of sail; it was the end of wood; it was the beginning of steam-driven cast-iron battleships with revolving gun turrets. The North immediately built six more Monitors and the *Monitor* became one of the key vessels of the US Navy. It was copied by Britain and there was soon 'no doubt that the Monitor was the progenitor of all the turreted vessels in the fleets of the world'.[69] After Hampton Roads, Ericsson saw the implications. He wrote to President Lincoln:

Our cause will have to be sustained not by numbers, but by superior weapons. By a proper application of mechanical devices . . . if you apply our mechanical resources to the fullest extent, you can destroy the enemy without enlisting another man.[2]

It was the start of mechanised warfare; the familiar changeover from men to machines translated into military terms. And, with the introduction of machines, the introduction of the arms race and the importance of keeping one technological step ahead of the enemy. The *Merrimac* had four months of unchallenged supremacy before she was superseded by the *Monitor*. Weapon, counter-weapon, and (as the South tried torpedoes and primitive submarines) counter-counter weapon in the now familiar pattern of escalation.

The first submarine in the Confederacy was *The Pioneer* built by McClintock and Watson of New Orleans in 1861. The first submarine in America was *The Turtle*—built by David Bushnell almost a century earlier.

This vessel, operated by a Sargeant Esra Lee, was employed in an attempt in 1776 or thereabouts, on the *Eagle*, an English Man-of-War, which proved unsuccessful, owing to the sergeant not being thoroughly versed in the management of his curious craft. She was soon afterwards sunk in the Hudson River but was subsequently recovered by the inventor, though never used again.[70]

The Turtle was built of wood, had glass windows and was just large enough to accommodate one man. With a manually operated propeller it could reach a top speed of 3 mph. It submerged when the operator touched a foot valve to let in water, and resurfaced when pumps forced the water out again. It had 'air sufficient to support him thirty minutes', and was not a machine for claustrophobiacs.

Fixed to the stern of the submarine was Bushnell's 'torpedo'. It consisted of a powder magazine that could be detached and jammed into the side of an enemy ship by an iron screw. The magazine, its time mechanism pre-set, would explode after the submarine had made its getaway. Fortunately for the British, the *Eagle*'s copper hull was too thick to be pierced. But while the plan failed, the concept was a success. It 'proved practically that a charge of gunpowder could be fired underwater, which is incontestably the essence of submarine warfare'.[70] Like Stevens's floating fort, Bushnell's submarine was well in advance of its time. The government refused to back it.

America's second submarine was built for France in 1801: the *Nautilus*, which its maker Robert Fulton called his 'plunging boat'. Fulton (1765–1815), a contemporary of Whitney, Lowell, Oliver Evans and his rival John Stevens, was a giant of steam and a giant of invention. Like Morse, he was a painter turned engineer.

Fulton was, if anything, cosmopolitan; born a British subject in a

Robert Fulton, 1765–1815

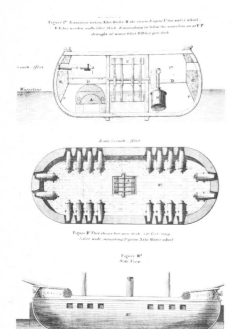

Fulton's 'Demologos'

British colony, the seed-time of his life was spent in England, the fruition took place in France and the harvest was reaped in his native America.[71]

Fulton's popular claim to fame rests on the *Demologos* and the *Clermont*. The *Demogolos* was the steam-driven floating fort that had been preferred to Stevens's design in 1811 and was the prototype of the *Merrimac*. The *Clermont* was his first steamboat, built in 1806. Nicknamed 'Fulton's Folly', the boat's successful voyage up the Hudson river ushered in the era of steamboat navigation on America's inland waterways. It ceased to be a folly. But it was prior to this (during his years in England and France) that Fulton, applying himself in the uncharted depths of underwater combat, revealed a vision probably unparalleled in the annals of war. In England he learnt his trade as a canal engineer. In France, during the Napoleonic wars, he turned against his former hosts and began working on a submarine for the French Navy. His motives were libertarian:

It is the naval force of England that is the source of all the incalculable horrors that are committed daily. . . . If by means of the 'Nautilus' one could succeed in destroying the English Navy, it would be possible with a fleet of Nautilus' to blockade the Thames to the end that England would become a republic . . . this would be a long step toward liberty and peace.[70]

It was an idealism shared by Beethoven and Wordsworth.

Apart from its small conning tower, the *Nautilus* looked like an ordinary sailing dinghy. But when it was under water, sails were stowed away and the mast slotted into a groove in the deck. Like Bushnell's prototype, the vessel submerged and resurfaced by letting in or pumping out water from tanks in the hull. It too

The 'Clermont', Fulton's first steamboat

was hand-driven, but this time there was space for four men inside instead of one.

It never saw action, but Fulton records his war games experiments in his own book, *Torpedo War and Submarine Explosions*, published in 1810. His object was to convince Congress of 'the practicability of destroying ships of war by this means'. He describes the demonstration he gave in 1805 'to convince Mr Pitt and Lord Melville that a vessel could be destroyed by the explosion of a torpedo under her bottom'. Fulton, disillusioned with Napoleon's imperial ambition, was back in England, with a new allegiance for his torpedo effort.

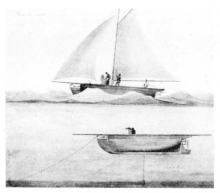

Fulton's painting of his first submarine the 'Nautilus'

At forty minutes past four the boats rowed towards the brigg [The *Dorothea*] and the torpedoes were thrown into the water: the tide carried them . . . under the bottom of the brig, where, at the expiration of eighteen minutes, the explosion appeared to raise her bodily about six feet: she separated in the middle, and the two ends went down; in twenty seconds, nothing was to be seen of her except floating fragments.[72]

Fulton commented that 'the experiment was of the most satisfactory kind', but Nelson's victory at Trafalgar put an end to Fulton's contract with the British Navy.

Fulton's torpedo was simply his own version of an underwater clockwork bomb: 'A copper case two feet long, twelve inches in diameter, capable of containing one hundred pounds of powder'. It was cased in cork to keep it buoyant, and on top of the case a small box with a gun lock was primed to set off a charge. It could be floated against a vessel, or anchored to the sea-bed and fitted with a tripping lever:

'A view of the brig "Dorothea", as she was blown up on the 15th of Oct. 1805'—from Fulton's 'Torpedo War and Submarine Explosions'

It is obvious that if a ship in sailing should strike the lever, the explosion would be instantaneous, and she be immediately destroyed; hence, to defend our bays or harbours, let a hundred, or more if necessary, of these engines be anchored in the channel, for example, the Narrows, to defend New York.[72]

'The anchored Torpedo, so arranged as to blow up a vessel which should run against it'

If Fulton's advice had been taken, the Yankees needn't have feared the *Merrimac*. Fulton calculated the cost: 1400 anchoring torpedoes at $84 each would cost $117,000; 1300 clockwork torpedoes, more expensive at $150, would cost $195,000.

The consideration which will now present itself, is, that the enemy might send out boats to sweep for and destroy the Torpedoes . . . (but) it amounts to an impossibility for an enemy to clear a channel of Torpedoes, provided it were reasonably guarded by land batteries and row boats. Added to the opposition which might be made to the enemy, there is great difficulty in clearing a channel of Torpedoes with

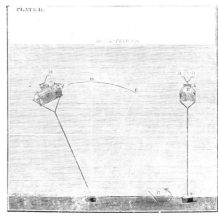

Fulton's harpoon gun positioned on the stern of the rowing-boat and attached to a Torpedo. Once the harpoon has been fixed on the bow of the enemy ship, 'it will then only be necessary to row away'. The line attached to the ship will pull the Torpedo from the rowing boat, setting off the clockwork timing device. The ship's motion will then pull the Torpedo towards it.

any kind of sweep or drag, so as to establish full confidence in sailing through it. It is only they who put down and know the number, that could tell when all were taken up.[72]

If the anchored torpedo is the forerunner of the modern mine, the clockwork harpoon is the real forerunner of the modern torpedo. An incredibly ingenious device—a small floating box that, when struck by a passing ship, causes an explosion in a mine anchored on the sea-bed, releasing a harpoon that shoots upwards through the water into the 'larboard and starboard bow'.

My experience with this kind of harpoon and gun, is, that I have harpooned a target of six feet square fifteen or twenty times, at the distance of from thirty to fifty feet, never missing, and always driving the barbed point through three inch boards up to the eye, which practice was so satisfactory, that I did not consider it necessary to repeat it.[72]

A variation on the theme, probably more economical, was to place the harpoon on a rowing-boat.

The man who shall be stationed at the gun, and who may be called the harpooner, is to steer the boat and fire when sufficiently near. If he fixes his harpoon in the bow of the enemy, it will then only be necessary to row away . . . should the harpooner miss the ship, he can save his Torpedo and return to the attack.[72]

The harpoon was attached to the rowing-boat by a line. Fulton argues that rowing-boats at night have an advantage of stealth, numbers and manoeuvreability. Put fifty boats against one man-of-war: 'Are not the chances fifty to one against the ship?' One harpoon would be bound to hit home, and 'it shall hereafter be no dishonour for a ship of the line to strike her colours and tamely submit to superior science and tactics'.

Fulton squared his conscience:

Then may not science in her progress, point out a means by which the application of the violent, explosive force of gun-powder shall destroy ships of war, and give to the seas the liberty which shall secure perpetual peace between nations that are separated by the ocean?[72]

And 'On the imaginary inhumanity of Torpedo War':

Men . . . exclaim that it is barbarous to blow up a ship with all her crew. This I admit, and lament that it should be necessary; but all wars are barbarous, and particularly wars of offence. It is barbarous for a ship to fire into a peaceable merchant vessel, kill part of her people, take her and the property, and reduce the proprietor with all his family from affluence to penury. It was barbarous to bombard Copenhagen, set fire to the city, and destroy innocent women and children. It would be barbarous for ships of war to enter the harbour of New-York, fire on the city, destroy property, and murder the peaceable inhabitants; yet we have great reason to expect such a scene of barbarism and distress, unless means are taken to prevent it; therefore, if Torpedoes should prevent such acts of violence, the invention must be humane.[72]

It was a scientist's argument that was to be used increasingly as the destructive power of science increased. But Fulton was denied the opportunity of putting his ideas into practice. The US Navy said that his torpedoes were 'comparatively of no importance at all'.[70] Not until sixty years later were the Confederates to turn to Fulton's ideas and designs, out of desperation. That the *Nautilus* was the model for McClintock and Watson's *Pioneer* shows how underwater technology had stagnated for half a century.

The *Pioneer*, privately financed, was begun in New Orleans in 1861 and completed the following year. She weighed four tons, had a crew of three and carried a 'magazine of powder'. On trial she succeeded in blowing up a target barge, but on her first voyage out she sank, like so many of the early submarines, with all hands. But, as McClintock wrote, 'this boat demonstrated to us the fact that we could construct a Boat, that would move at will in any direction desired, and at any distance from the surface'.[73] Accidentally dredged from the river-bed sixteen years later, the *Pioneer* was presented to the City of New Orleans and, in 1910, fixed in concrete as an historical monument.

A second submarine made its contribution to engineering in that it was originally designed to be propelled by an engine instead of by men. But it too sank, and the first McClintock and Watson submarine to see action was named after its inventor and the third member of the firm, the *Horace L. Hunley*. Twenty-five feet long, four feet in diameter and driven by

The 'Horace L. Hunley' at Charleston, South Carolina, 6 December 1863, two months before her destruction. Clearly visible are the spar for attaching the torpedo, side fins, hatches, and propeller. The man leaning on the rudder shows how impossibly cramped were the conditions for the ten-man crew inside.

Confederate barrel torpedoes designed to destroy the United States flotilla in the Potomac River. The barrels supported cylinders of gunpowder which were ignited by a slow fuse after they'd wrapped themselves round an anchored vessel.

eight men, she was bigger and faster than her two predecessors. But her trials were not auspicious: 'She lacked longitudinal stability, and had a bad way of diving unexpectedly nose-on into the muddy bottom and sticking there until her crew suffocated.'[73] She did this four times and drowned her inventor. Her death toll during trials was thirty-five and she was christened 'The Peripatetic Coffin'.

The 'coffin' carried her torpedo attached to her bow by a spar. The idea, considerably more primitive than Fulton's, was to ram the torpedo into the side of the enemy and then reverse before the explosion. On the night of 17 February 1864, she came into contact with the USS *Housatanic*. Manoeuvred alongside, the *Horace L. Hunley* became the first American submarine to sink an enemy ship. Unfortunately, the *Hunley* didn't reverse fast enough and was blown up with her victim. It was the end of Confederate submarines.

In place of the submarine—the torpedo boat. Both sides, learning from Fulton's experiments with harpoon guns on rowing boats, built reasonably effective torpedo boats. Most of them were variations on the Fulton theme—submerged gunpowder kegs activated by wheels and levers as ships passed overhead. The exception was the electric mine, the invention of Samuel Colt, which was activated by an operator passing an electrical impulse along an underwater cable from his concealed position on shore.

In June 1841, twenty years before the outbreak of the Civil War, Colt had written to the President:

By referring to the Navy State Papers, page 211, you will discover that Robert Fulton made experiments which proved that a certain quantity of Gun-Powder discharged under the bottom of a ship would produce her instant destruction. . . . Discoveries since Fulton's time combined with an invention original with myself, enable me to effect the instant destruction of either Ships or Steamers, at my pleasure on their entering harbour, whether singly or in whole fleets; while these vessels which I am disposed to allow a passage are secure from the possibility of being injured. All this I can do in perfect security and without giving an invading enemy the slightest sign of his danger.[36]

In June 1842, Colt proved his point:

I made an experiment with my submarine battery from on board the US '74 the *North Carolina*, on a vessel about one hundred tons which was being towed through the water at the rate of about three knots an hour. Many of the Navy officers witnessed the effects from on board the *North Carolina* and seemed satisfied the power employed was sufficient to have destroyed a vessel of almost any tonnage.[36]

And, on 14 August:

On this occasion, I operated from Alexandria, a distance of 5 miles, & blew the vessel to pieces so quick after the signal was given that the noise of the explosion and the report of the signal gun was blended together in one sound.[36]

The USS 'Commodore Barney' was damaged by an electrically-fired mine in the James River, Virginia, on 5 August 1863

The electric cable used by Colt was insulated by cotton yarn, waxed and enclosed in metal. Supporting the young inventor, for reasons of his own, was Samuel Morse, who could see in the cable a way for the telegraph to cross water. Despite Colt's success, the US Navy declined to mine its harbours. Experimentation ceased and Colt returned to make his fortune with the revolver.

It was the Confederates who mined the harbours. In 1864, in its attempt to take Richmond, the Confederate capital, the Union Navy raided the James River. Mine-sweeping techniques, despite Fulton's predictions, easily dealt with most of the mines that stretched across the river mouth. Only the shore-operated electric mines lay between the Union and Richmond. As the Union vessels slowly advanced upstream, the mines were exploded and the *Commodore Jones* blown out of the water. Colt's boast of 1841 had come true, and Richmond was saved.

Neither the torpedoes nor the ironclads were restricted to the sea. The land-going versions of the *Merrimac* were the iron-plated railroad cars used by the Union to rebuild bridges destroyed by the Confederates. Used on the Baltimore and Ohio, they were the precursors of the armoured car and—precedents by Leonardo da Vinci aside—of the tank. The land-going versions of the torpedoes were the buried gunpowder kegs that were the world's first landmines. Introduced by the South in 1862, they were denounced as inhuman, but were soon used by both sides, most spectacularly in 1864 when the Union, in an attempt to take Petersburg, employed the tactics of the medieval castle builder. Tunnelling for 500 feet, they arrived beneath a Confederate fort, part of the city's defensive outworks. Instead of the medieval pit props and combustibles burnt so that the tunnel would collapse and the castle with it, the Union placed, in each of eight chambers, a ton of gunpowder. At 5.00 am on 30 July, the mine was fired. It blew a huge crater in the ground, the fort disappeared, and the Confederate line gaped a quarter mile wide. That the Union failed to take advantage of the Battle of the Crater was the fault of generalship, not gunpowder.

Military innovations on land, sea and, finally, in the air. Balloons had been used by Napoleon's forces for aerial recon-

Union Parrot gun mounted on a specially constructed railroad car, Petersburg, Virginia, 1864

naissance in 1794. They were introduced by the Union Army in 1861 for the same purpose. Chief of the Union's Aeronautic Section was Professor Thaddeus Sobieski Coulincourt Lowe—encouraged by the irrepressible Joseph Henry. Lowe's use of the balloon was revolutionary: he was the first man to use the camera to photograph enemy positions from the air, and the first man to use the telegraph for air to ground communication. He gave a demonstration at the White House and recorded his observations for posterity and President Lincoln.

Thaddeus Lowe, 1832–1913

To the President of the United States.
This point of observation commands an area near fifty miles in diameter—the city with its girdle of encampments present a superb scene —I have pleasure in sending you this first despatch ever telegraphed from an aerial station and in acknowledging the indebtedness to your encouragement for the opportunity of demanding the availability of the science of aeronautics in the military service of the country.

At the Battle of Fredericksburg in 1862, Lowe first spied out the land: 'I took observations to ascertain the best location for crossing the Chickahominy River.'[1] Next day:

. . . with my telegraph cables and instruments, I ascended to the height desired and remained there constantly during the battle, keeping the wires hot with information.[1]

Confederate preparations for a forthcoming offensive were spotted at once by Lowe, and the Union Army planned counter-offensive action. The Confederates tried to shoot Lowe down. Failing, they resorted to other tactics.

The General commanding this Army Corps wishes every precaution taken to prevent the enemy from discovering by balloons or other means, the numbers of our advanced outposts. No lights should be kept at night except where absolutely necessary, and then under such screens as may conceal the lights from observation. Further, tents, if used, ought to be pitched under cover of woods and sheltered in all cases as far as possible from accurate computation.[75]

Camouflage, dummy guns—the whole operation was like a dry-run for the aeroplanes, black-out and camouflage of World War II.

Lowe's novel use of the telegraph in the air was paralleled by its use on the ground. The field telegraph, devised by the Union, consisted of portable telegraph apparatus for transmitting and receiving, and ten miles of wire to be strung up where and when needed. The horse-delivered message was replaced by instantaneous communication. The Union Army, in five years, erected 15,000 miles of cable and sent, reportedly, six million messages.

Politicians could control generals, generals could control armies. For the first time, power in wartime could be centralised from positions hundreds of miles behind the front lines.

A less obvious use of the balloon and the telegraph was in collecting and reporting weather observations. Franklin had shown great interest in meteorology, and in 1785 American balloonist John Jeffries had carried instruments over the English Channel in an attempt to study wind patterns. By 1854, encouraged yet again by Joseph Henry, the Smithsonian Institution had weather observers in thirty-one states, all sending in their reports by telegraph. By 1861 there were 400 stations. Weather reports were regarded as essential military intelligence and have been ever since.

Surprisingly, the one thing balloons were not used for was aerial bombardment, though the idea had been suggested during the war with Mexico. It was another twenty years before Captain Carl Peterson patented an 'aerial war vessel' 'which may be used for dropping bombs to cause explosions and fires in enemy territory'.[76] It consisted of a series of balloons propelled by an air screw and joined together to form a long sausage. The idea, figuratively or literally, did not get off the ground—yet.

Several inventions of an even less practicable nature had their origins in the Civil War. The 'Improved Swimming Device', illustrated in *Scientific American* in 1880,[76] utilised the wheels, crankshaft and propeller of the manually operated submarine. The man at the controls drove the machine through the water, without the protection of a hull. The Natural Flying Machine of 1865 was proposed in a letter to a Baltimore newspaper, and was to consist of a circular metal frame in the middle of which was a

Lowe, about to cast off

Lowe's balloon 'Intrepid', ready to go up on a reconnaissance mission. The problem with balloons was filling them. Hydrogen gas had to be specially made.

*City Point, Virginia, 1864—in the fore-
ground big Union mortars to be used in the
siege of Petersburg later that year, on the
beach a line of 30-pound Parrot guns*

*Union Cavalry Private equipped with
sabre and pair of Colt Navy revolvers*

*Private Robert Patterson, 12th Tennessee
Infantry. For the first time uniforms were
mass-produced.*

*Confederate dead, Spotsylvania Court
House, Virginia, May 1864*

Union troops in the trenches outside Petersburg, Virginia, March 1865. Two weeks later the city fell and the war ended.

basket for the 'pilot'. The machine was to fly by the natural action of a dozen American eagles wearing jackets which tethered them to the frame. Whether or not practicable, or even serious, the suggestions show a continuing interest in both the submarine and ballooning even after the military had declared both ideas dead. The Union grounded its balloons after 1863.

On 3 April 1865, Petersburg finally fell to the Union Army. On 9 April, the surviving Confederate forces surrendered. The American Civil War was over. Machinery had won. Half a million men died. It left the South in need of reconstruction and with the problem of coming to terms with an industrial society and a new way of life. It left the North intact and looking forward to a bright industrial future. It left the United States with one of the best-equipped and tactically most experienced armies and navies in the world—a position it never lost. But, of all the inventions that came out of the Civil War, the most significant was the invention of the modern war itself.

Chapter 4

The United Nation

AMERICAN OLEOGRAPH CO. MILWAUKEE

Philadelphia, 10 May, 1876. Ten o'clock. President Ulysses S. Grant opens the great Centennial Exhibition:

One hundred years ago, our country was new and but partially settled. Our necessities have compelled us chiefly to spend our means and time in felling forests, subduing prairies, building dwellings, factories, ships, roads, machinery. . . .

Burdened by the great primal works of necessity, which could not be delayed, we have yet done what this exhibition will show in the direction of rivalling older and more advanced nations.[77]

The scars of the Civil War were beginning to heal. The nation was again united. It met together to celebrate a century of discovery; of conquest of the alien environment, and of heroes of invention who had made it possible. Franklin, Fulton, Howe, Morse —their faces appeared on the certificates of stock issued to float the exhibition.

The Centennial was the visual fulfilment of the American Dream: 'convincing evidence of the dauntless spirit which animates our citizens, and the determination to succeed which is one of their chief characteristics'.[79] 284 acres of grounds, 249 buildings, and 510,000 square feet of machines in Machinery Hall convinced the Americans of their achievement. Longfellow saw 'all the wonders'. Emerson, now seventy-three, was 'dazzled and astounded'.[77] There was every kind of machine—milling, grinding, sewing, screwing, printing, casting, punching, polishing. There were drills, hammers, mills, engines, presses and pumps. And seventy-five locomotives.

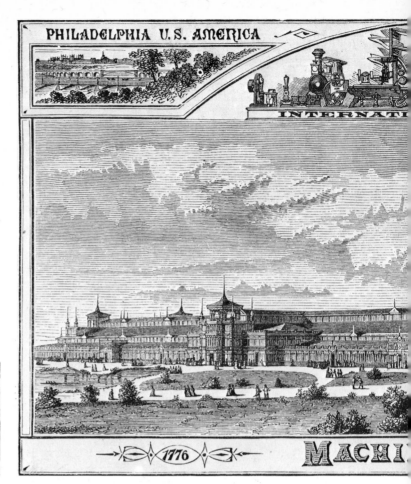

Lyall's Positive Motion Looms

B. Grant's Calculating Machine

Something at once sublime and infernal in the spectacle. Machines claim nothing for themselves; they make no boasts, but silently perform their task before your eyes; the mode in which it is effected is a mystery; the spools, spindles, shuttles are there, so is the raw material. One sees the means and the result, but the process is invisible and inscrutable as those of Nature. . . . Nowhere else are the triumphs of ingenuity, marvels of skill, so displayed, so demonstrated. [*Atlantic Monthly*, July 1876]

Bishop Simpson specifically remembered the machines in his Inaugural Prayers:

We thank thee God for social and national prosperity and progress, for valuable discoveries and multiple inventions, for labour-saving machinery relieving the toiling masses.[77]

Machinery had shouldered the burden, liberated the people. Mass-production had made the good life possible.

In going through the department of Home Manufactures, I was constantly struck by the better provisions made for the millions than the few. [*Atlantic Monthly*, October 1876]

A good life possible for everyone, not just the rich—and America no longer the poor relation of Europe.

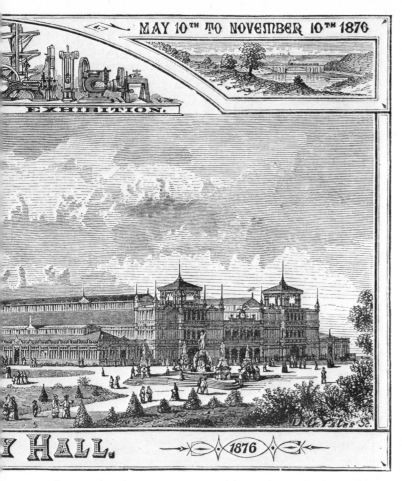

MAY 10TH TO NOVEMBER 10TH 1876

EXHIBITION.

Y HALL. 1876

Howes, Babcock and Co's Grain Cleaning Machinery

Ferracute Machine Works' Presses and Dies

. . . the superior elegance, aptness and ingenuity . . . is observable at a glance [and] all that Great Britain and Germany have sent us is insignificant compared with our own contribution.

Britain displayed one cotton machine: sewing cotton by John Dewhurst and Sons ('remarkable for its strength, pliability, evenness and freedom from knots'), and the hand-made Patent Treble Wedge-fast Breechloader by W. W. Greener of Birmingham ('the strongest and most durable breechloader ever invented'). Water closets, valentine cards, a three-wheeled pram and, amongst 3000 other items, a view of Palestine. Once bitten, twice shy:

The tariff is too high to try to sell in the US and it would be foolish to let Americans see our machinery on show for six months.[77]

Nevertheless,

It would be foolish not to recognise the fact that Great Britain was in face of her most powerful rival in manufacture.[77]

Customary British scorn, from *The Times*, turned to admiration:

The American invents as the Greek sculpted as the Italian painted. It is genius.[77]

The Centennial Exhibition was America's first World's Fair and the first occasion on which foreigners came to the United States to learn. Fifty years before, the Atlantic traffic had been entirely in the other direction.

In the week that the Exhibition opened, celebrating the triumph of man, machines and civilisation, the Wilderness—and Little Big Horn—suddenly intruded:

A sorrowful piece of news is that which tells us of the slaughter in a fight with the hostile Sioux, of General Custer and all the officers and men of five companies of cavalry. Into the jaws of death rode the three hundred. This intelligence has shocked the whole country and had it been received three days earlier must have cast a shadow of gloom over the otherwise joyous celebrations of the Centennial Anniversary. [*Philadelphia Public Ledger*]

The death of Custer served as a reminder that parts of the US were still frontier country: white man's technology was still fallible. But fallibility was there to be overcome. Proud of its past, America was confident of its future—and in that future there was no doubt as to the role of technology.

Even in some of the minor Centennial exhibits, there were indications of what was to come. The Mechanical Orchestrian anticipated Muzak; the clean-lined coffee urn the functionalism of twentieth-century design. 'Elegant' mass-produced furniture demonstrated that factory production and good taste were not incompatible. Special demonstrations showed that mass-production could be applied to such an intricate and hitherto entirely hand-made object as a watch. And if to watches—then to anything. Even to what Mark Twain called 'this curiosity-breeding little joker'—the typewriter.

The typewriter began with a British patent of 1714 which described

an artificial machine or method for the impressing or transcribing of letters singly or progressively one after another, as in writing, whereby all writings whatsoever may be engrossed in paper or parchment so neat and exact as not to be distinguishable from print.[80]

Nothing else is known of this machine or whether, in fact, it ever became a machine, and the first modern typewriter was patented in America in 1867 by C. L. Sholes. It had evolved from the search for a printing device for the telegraph—Morse had invented a keyboard in 1845. Sholes, a printer, had been developing a machine that would print numbers on tickets. 'Why cannot the paging machine be made to write letters and words and not figures only?' suggested Sholes's partner Glidden. Between them, they'd conceived the typewriter.

A young woman working at a Sholes' typewriter in 1872. The machine was so designed that it was virtually impossible for her to see what she was typing.

The first commercially sold typewriter, the Remington Model 1. But no one wanted to buy it.

The early models, based on numerous European precedents, had the familiar travelling carriage, key bars, arranged in a semi-circular basket and which hit the roller with an upward strike, and a keyboard. The letters, based on an analysis of the frequency of their use in the English language, were positioned on the keyboard so that the key-bars were least likely to collide when they moved upwards.

Philo Remington bought the patent from Sholes in 1873. His small-arms factory, feeling the effects of a market slump after the Civil War, was already manufacturing sewing-machines, and Remington believed that, like the sewing-machine, the typewriter was ideal for mass-production. It was. But he also needed a market, and there wasn't one. Despite Mark Twain's commercial and a Centennial 'come-on'—25 cents to have your message typed and sent home—the typewriter was a flop. Advertised as the successor to the pen—a device enabling 'children, blind and aged [to] print at once'—it was dismissed as an 'aid for the myopic' and 'a prop for sufferers of writer's cramp'.

Society sometimes needs time to catch up with technology—to know what to do with an invention. Only in the late 1880s, when technical improvements made possible speeds of up to ninety words a minute, did the machine begin to catch on—in business offices, for writing letters, a use that Sholes had never even anticipated. And, in the process, it struck one of the first blows for Women's Lib. As female typists invaded offices, the male clerk was ousted. 'Respectable' girls found one of the first 'respectable' job outlets outside the home.

The Centennial Committee gave 12,000 awards: to established heroes like Colt, Westinghouse, Pullman, McCormick; and to the less well-established—Du Pont for gunpowder, Chessborough for Vaseline, Otis for the elevator (English read lift). And to unknowns—like Alexander Graham Bell.

Philo Remington, 1816–1889

The Remington typewriter works in 1901—fifteen years after Remington had been forced to sell the business as a flop

LAWYER   Now I am going down town to my other office—and I want you to keep me posted by speaking through the telephone.

DERBY (*a negro clerk*)   The what?

LAWYER   The Telephone. I had it placed in here yesterday, and of course you don't understand it. Here it is. (*Going to back flat and showing it to Derby.*)

DERBY   That fish horn?

LAWYER   A fish horn. No sir—the greatest invention of the nineteenth century.

DERBY   Oh.

LAWYER   You see Derby—this telephone connects with my down town office and all I have to do when I'm sitting in my other office down town—to convey my message to you—I speak through this telephone—then you answer me back—I can hear you and you can hear me as plain as you do now—so you will understand—should anyone call—speak through the pipe and let me know—I shall be at the other end and tell you what to do.

DERBY   Am I to blow through it?

LAWYER   Blow through it—No sir—Speak through it—you understand—Remember my instructions. (*Exit.*)

DERBY   Lawyer Taylor—going crazy. Speak through that funnel—and he hear me at the Battery. No wonder he's baldheaded.[81]

Whilst the typewriter languished, the telephone didn't. Ed Harrigan played in his sketch *The Telephone* in November 1877— Harrigan and Hart and the 'Theatre Comique' were to New York in the 1870s what Radio City Music Hall is today. That the sketch appeared just eighteen months after Bell's demonstration at the Centennial shows not only that the telephone was an object of curiosity and fun but that it caught the public imagination and was put into commercial use faster than anything else in the history of American invention.

The latest American humbug—far inferior to the well-established speaking tubes,

said *The Times*, scornful again. Making British apologies at the turn of the century, Arnold Bennett dubbed it 'the greatest achievement of the American people', which was generous, as Bell, when he invented the phone, was still British—one of the still considerable number of Europeans crossing the Atlantic in search of creative outlets in the American ethos.

Bell had arrived in Canada at the age of twenty-three. A year later, he accepted a job at the Boston School for the Deaf. At the same time, the Western Union Telegraph Company were looking for a system whereby they could use one telegraph wire to transmit several messages at once. It was cheaper than erecting more

Alexander Graham Bell, 1847–1922

telegraph poles—and they offered a prize. Bell, already experimenting with the telegraph, took up the challenge: 'If I can make the deaf talk—I can make iron talk.'

Bell knew, from past experience and careful observation, that striking a note on a piano would cause a wire tuned to the same pitch on *another* piano to vibrate. The same sympathetic reaction could be achieved with tuning forks. Was this the solution to the 'multiple telegraph' problem? To send messages at different pitch? As many messages as there were notes on the piano? Bell's study of the membrane of the human ear enabled him to take his theory one step further.

It would be possible to transmit sounds of ANY SORT if we could occasion a variation in the intensity of an electric current like that occurring in the density of the air.[1]

The concept of the electrical transmission of human speech—of the telephone—was born. The practice took a little more time. Bell wrote to his parents in May 1875:

My inexperience in such matters is a great drawback. However, Morse conquered his electrical difficulties although he was only a painter, and I don't intend to give in either till all is completed.

Morse was dead. But the man who had helped and encouraged him was still alive: Joseph Henry, thirty years on, played the same role again:

He said he thought it was 'the germ of a great invention'. . . . I said that I recognised the fact that there were mechanical difficulties . . . I added that I felt that I had not the electrical knowledge necessary to overcome the difficulties. His laconic answer was 'Get It'. I cannot tell you how much these two words have encouraged me.[83]

Bell achieved his breakthrough in June.

His experimental equipment consisted of three telegraphic transmitters and receivers, joined by wires and placed in adjacent rooms. Each piece of apparatus contained thin metal strips of different lengths which Bell called 'reeds'. These were loose at one end and therefore free to vibrate, in the manner of the tuning fork or piano wire. Comes the classic 'discovery' situation. The accident. Bell's assistant, Watson, checking a contact point on his transmitter, adjusted a reed. Bell, in the next room, heard the faint 'twang' over the receiver. Immediately he recognised its significance. The first time 'a human ear heard the tones and overtones of a sound transmitted by electricity'.[2]

The vibrating reeds were no longer being used, as in all previous telegraphs, to make or break an electrical circuit. Now, without

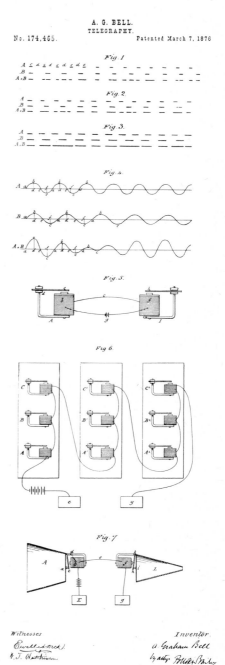

A. G. BELL.
TELEGRAPHY.
No. 174,465. Patented March 7, 1876

direct contact, they transferred their vibrations to the magnetic field surrounding the electromagnet, causing fluctuations in the intensity of the electric current passing down the wire. At the opposite end, the metal reed, acting as a primitive receiver, turned the fluctuating current back into vibrations—and sound. The revelation—and the revolution—was that in all previous telegraphs the electric current was either on or off: now it was always on but was capable of 'transmitting sounds telegraphically . . . by causing electrical undulation similar in form to the vibrations of the air'.[2]

Two days later, on 4 June 1875, Watson constructed the first primitive 'harmonic telegraph':

An instrument modelled after the human ear, by means of which I hope to transmit a vocal sound.[83]

Whilst capable of transmitting sounds, the embryonic telephone had not yet transmitted human speech:

I am like a man in a fog who is sure of his latitude and longitude. I know that I am close to the land for which I am bound and when the fog lifts I shall see it right before me.[83]

On 10 March 1876, three days after his patent was granted, the fog finally lifted and Bell succeeded in transmitting the first 'complete sentence of human speech'. The epoch-making words were: 'Mr Watson. Come here. I want to see you.' Sadly for the neatness of the story, these words were not transmitted with the aid of his patented machine but by another device, which varied the resistance by using a platinum needle which was attached to the transmitter diaphragm and dipped into an acid solution. Nevertheless, the magnetophone found its voice on 27 March, when Bell addressed his father with the word 'Papa'. Two months later, the phone was on show at the Centennial, where it elicited a reaction from the Emperor of Brazil—'My God, it talks'—that has passed into scientific folklore. But not fact. Sad again, for the neatness of the legend, these words were never uttered but have been traced to their source in a juvenile book published in 1923. The author, a Brooklyn teacher, explained 'that he wanted to stimulate pupils' interest in science by telling them colourful stories'.[77]

The reality is Bell's own prosaic account to his parents: 'Sir William [Thomson] and Dom Pedro then came to see my apparatus.'[83] Thomson was the Scottish scientist whose enthusiasm had been behind the successful laying of the first transatlantic telegraph cable in 1866. He was on the judging panel of the Centennial and, in his official report, was anything but prosaic.

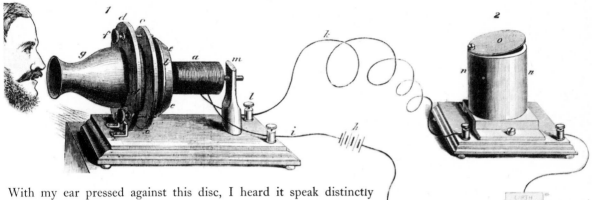

Bell's first telephone, 1876, as shown at the Philadelphia Exhibition. On the left is the transmitter, on the right the receiver.

With my ear pressed against this disc, I heard it speak distinctly several sentences . . . I need scarcely say I was astonished and delighted. . . . This, perhaps the greatest marvel hitherto achieved by the electric telegraph, has been obtained by appliances of quite a homespun and rudimentary character. With somewhat more advanced plans, and more powerful apparatus, we may confidently expect that Mr Bell will give us the means of making voice and spoken words audible through the electric wire to an ear hundreds of miles distant.

Six months after the exhibition, Bell gave his apparatus the more practical shape and greater power that it required. *Scientific American* reported:

This new apparatus no longer requires an electrical battery and consists of a powerful horse-shoe type magnet, the arms of which are provided with wire coils. This magnet is placed before a thin, soft-iron plate.[76]

The soft-iron plate was fitted into a mouthpiece and there responded to the vibrations of the voice of the speaker.

In the Spring of 1877, Bell demonstrated his new 'tele-phone' in public. On 12 February in Boston, Watson spoke into the transmitter. Fourteen miles away, in Salem, Bell, before an invited audience, spoke back.

As I placed my mouth to the instrument it seemed as if an electric thrill went through the audience, and that they recognised for the first time what was meant by the telephone.[83]

Sceptical journalists, watching in both towns, later compared notes. There was no fraud. America was agog. Of all the new-fangled devices she was getting used to, this was the *most* fangled. Reaction was the same in Britain, where Queen Victoria wrote in her diary for July, 'A professor Bell explained the whole process which is most extraordinary'.

For $100,000 Bell offered his patent to Western Union. They refused. They'd recently had their 'multiple telegraph' problem

Notice

Visitors are requested to abstain from conversation during the lecture in Salem, as slight noises in this laboratory may seriously interfere with the success of the experiments.

A. Graham Bell

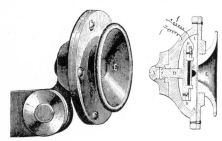

Edison's telephone transmitter

solved for them by Thomas Edison. Bell—a master of shrewd promotion—formed the Bell Telephone Association and continued his public demonstrations, charging $200 a lecture and packing the auditoriums. Western, realising they might have missed out on something, invited Edison to 're-invent' the telephone especially for them. He did. The major change was in the transmitter. Edison discarded Bell's magnet in favour of a carbon button, which still caused the vital fluctuations in the electric current but was able to improve the volume. Claimed Edison:

It will render audible in a large room, the conversation sent over a wire several hundred miles long, when the ordinary Bell receiver must be held close to the ear. This instrument will undoubtedly come into general use as soon as the numerous patents have expired.[85]

Western bought Edison's transmitter for the $100,000 they'd already refused Bell for the whole apparatus. And they didn't wait for the patents to expire. Nor did Bell. As Edison said, 'The Bell company in England were infringing my transmitter at the same time that I was infringing their receiver.'

Court action and counteraction for piracy were further complicated by the third telephone inventor: Emile Berliner. Months before Edison, Berliner had produced an Edison-type mouthpiece. Bell had bought the Berliner patent and used it as a stick with which to beat Edison and the Western Union. It left the world with a hybrid instrument: Edison mouthpiece, Bell receiver—a good compromise.

The box telephone—the first commercial telephone. It incorporates both the transmitter and receiver and gets rid of the battery, although the box had to be up to fourteen inches long to accommodate the large horseshoe magnet.

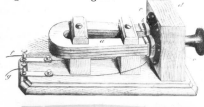

Elisha Gray had experimented with electromagnets and tuning forks at the same time as Bell and in much the same way. On the very day that Bell applied for his patent—14 February 1876—Gray filed a caveat with the Patent Office: 'I claim as my invention the art of transmitting vocal sounds or conversations telegraphically, through an electric circuit.'[83] But whereas Gray was giving notice of his intention to invent the telephone, Bell had already produced an instrument and won the acclaim. Gray, instead, made money by giving performances of his eventual invention, 'the musical telephone which is built upon the framework of a grand piano'.[76] The instrument, which looked like a cross between an organ and a megaphone, was adjudged to sound 'Loud and pleasant'. It was as popular, though with a different kind of audience, as Harrigan and Hart at the Theatre Comique.

The first telephones sold commercially were for private circuits. They were offered for $3 with a guaranteed range of one mile. For five miles you paid $5. But Bell, as early as 1877, envisaged something much bigger: a public service.

At the present time we have a perfect network of gas-pipes and water-pipes through our large cities. We have main pipes laid under the streets communicating by side pipes with various dwellings, enabling the members to draw their supplies of gas and water from a common source.

In a similar manner, it is conceivable that cables of telephone wires could be laid underground, or suspended overhead, communicating by branch wires with private dwellings, counting houses, shops, manufactories etc. etc., uniting them through the main cable with a central office where the wires could be connected as desired, establishing direct communication between any two places in the city. Such a plan as this, though impracticable at the present moment, will, I firmly believe, be the outcome of the introduction of the telephone to the public. Not only so, but I believe in the future, wires will unite the head offices of the Telephone Company in different cities, and a man in one part of the country may communicate by word of mouth with another in a distant place.

By 1879, New York possessed a central telephone exchange. The songs followed: 'Hello Central. Give Me Heaven' (1901). So did the subscribers—New York's hotels amongst the first. And so did Women's Lib, yet again. Out went the male operators, in came the girls, to fill the second 'respectable' job that society had offered within a decade.

The spread of the telephone was phenomenal. In 1880, instruments were already owned by 50,000 Americans. By 1900 there were three million owners, and a million miles of wire stretching across the country. Under President Taft, the telephone exchange penetrated the White House. By 1918, the United States possessed seventy-five per cent of the telephones in existence. New York alone had as many as Great Britain. Newspapers and businesses very quickly came to regard the phone as essential equipment. It was faster than the mail, more personal than the telegraph. More important, perhaps, was the way the phone continued the process of uniting rural and urban America, of unifying the sprawling Nation. It was soon regarded as the 'greatest civilising force we have, for civilisation is largely a matter of intercommunication'. The only civilised loss was the art of letter writing.

For Bell, commercial gain was subordinate. For Edison it was motivation:

The first person to publically exhibit a telephone for the transmission of articulate speech was Alexander Graham Bell. The first practical commercial telephone was invented by myself.

Bell always described his profession as 'teacher of the deaf'. He

The portable telephone—'which will carry messages five or six miles . . . It is preferable to employ two, as represented, to prevent the confusion frequently consequent upon the two persons conversing, when speaking at the same time, which is oftentimes the case where only one telephone at either end is used.'

The first commercial telephone switchboard, used in New Haven, Connecticut, in 1878, had eight lines and twenty-one subscribers

Girls operating a pyramid switchboard in 1881

In Boston, in 1883, the Tremont office still used some men on the switchboard. But the casual gent smoking a cigar on the left is the manager.

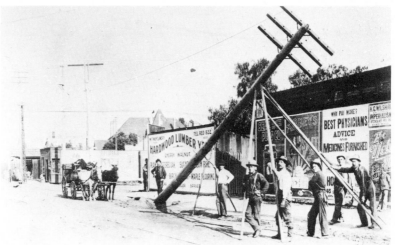

Construction crew celebrating the completion of a pole line in the big city—New York, 1890, population 3½ million

Telephone poles played havoc with the urban landscape—the highest pole line on West Street, New York, at the south-west corner of 66th Street and 11th Avenue

Construction crew in the small town—Redwood, California, 1900, population around 10,000

Alexander Graham Bell at the New York end of the transcontinental telephone line to Chicago, opened in 1892

was the inventor-amateur. Edison was the first full-time inventor-professional. A different breed. The inventor who turned the actual process of invention into Big Business.

If the Centennial was the visual fulfilment of the American Dream, Edison was its personification. A dropout from school at the age of twelve, he became a living legend: the archetype of the 'ordinary-boy-makes-good', the self-made American, the 'Wizard of Menlo Park'. With 1093 inventions to his name, the Colossus of Invention. And almost as many quotable aphorisms that reinforce the American image: 'There is no substitute for hard work'; 'Genius is 99 per cent perspiration and 1 per cent inspiration'.

Edison had a precocious schoolboy's interest in maths and chemistry—blithely lowering nitro-glycerine into the Boston sewers, and on another occasion, burning himself so badly he didn't go out for two weeks. He also had an early interest in electricity:

The King of Terrors in Edisonian garb was an unknown factor (in the calculations of cockroaches). Our ingenious trickster proceeded to secure certain alluring strips of shining tin foil against the office walls, baited with such edibles as appeal most strongly to the gastronomic instincts of the cockroach, connected the strips with a powerful battery, and was rewarded by the spectacle of a steady rain of calcined insects pouring from the improvised crematoria overhead.[85]

He had an even earlier interest in the Morse code and the telegraph.

He was always tinkering with telegraphy and once rigged up a line from his home to mine, a block away. I could not receive very well, and sometimes I would climb out, climb on the fence and holloa over to know what he said. That always angered him; he seemed to take it as a reflection upon his telegraph line.[85]

The passion continued. He even called his children 'Dot' and 'Dash'. And it was this interest in the telegraph that inspired his first inventions.

1869: The Improved Stock Ticker. Important, as it precipitated Edison's first excursion into the invention and manufacturing business. He employed fifty men and turned out stock-tickers from a factory in Newark, New Jersey, where, when the tickers developed a 'bug', he imprisoned his men for sixty hours till the fault was located and the tickers running smoothly again.

1872: The Duplex Telegraph. 'I experimented 22 nights—tried 23 Duplex systems, 9 failures, 4 partial success, 10 all right, 1 or 2 bad'.

1876: The Quadruplex. 'I have struck a new vein in Duplex Telegraphy'.[86]

Later, the Sextuplex. All successful attempts to solve Western Union's multiple-telegraph problem—the same problem that was exercising Bell. The Quadruplex, which enabled two messages to be sent in one direction and two in the other, at the same time, saved America $15 million. The President of Western Union, whose company made the most saving, called it, modestly in the circumstances, 'one of the most important inventions in telegraphy'. It was for another 'important telegraphic invention', the electromagnetic shunt (which momentarily reversed live current), that Edison received his award at the Centennial. He was twenty-nine.

The original tin-foil phonograph invented by Edison in 1877

From his work on the telegraph came the acoustic telegraph (his name for the telephone), the microphone, the megaphone and, most successful of all, the phonograph. In 1877, experimenting with a telegraph transcribing machine, Edison ran a needle over a telegraph tape indented with dots and dashes. It gave a 'light, musical, rhythmic sound resembling human talk heard indistinctly'. The same year, looking for a replacement for the iron membrane diaphragm of Bell's telephone, he experimented with short needles.

I discovered the principle by the merest accident. I was singing to the mouthpiece of a telephone when the vibrations of the voice sent the fine steel point into my finger. That set me thinking. If I could record the actions of the point and send the point over some surface afterward, I saw no reason why the thing would not talk. I tried the experiment first on a string of telegraph paper [waxed strips about half an inch wide] and found that the point made an alphabet [i.e. marked the paper with tiny holes]. I shouted the words 'Hallo Hallo!' into the mouthpiece, ran the paper back over the steel point and heard a faint 'Hallo Hallo' in return. I determined to make a machine that would work accurately and gave my assistants instructions, telling them what I had discovered. That's the whole story. The phonograph is the result of the pricking of a finger.[85]

Thomas Alva Edison at Matthew Brady's photographic studio in Washington, D.C., April 1878, with his improved tin-foil phonograph

Edison's first experimental machine worked on exactly the same principle except that the waxed telegraph paper was replaced by tin foil wrapped round a spirally grooved metal cylinder. As the cylinder was rotated, the grooves provided the continuous path for the needle to travel over; and the tin foil, the material on which to record the sounds. Into the horned mouthpiece that contained a needle attached to a diaphragm, Edison quoted the immortal words: 'Mary had a Little Lamb'. The needle vibrated along the grooves, indenting the tin foil. After the recording, the cylinder was stopped and wound back to the start point. Using a more sensitive needle and diaphragm, Edison traced the same path

Emile Berliner's gramophone

over the same grooves. Back it came, faintly: 'Mary had a Little Lamb'. It worked. It was immediately patented and became an equally immediate sensation. A machine that could record, store and reproduce the human voice.

In December 1877, representatives of *Scientific American* visited Edison's lab. 'The machine began by politely inquiring as to our health, asked how *we* like the phonograph, informed us that *it* was very well and bid us a cordial good night.'

With more Barnum and Bailey gimmicks, the instrument was demonstrated up and down the country, netting thousands of dollars a week, and Edison hurried into print with his suggestions for the machine's use.

1 Letter writing and all kinds of diction without the aid of a stenographer. 2 Phonographic books, which will speak to blind people without effort on their part. 3 The teaching of elocution. 4 Reproduction of music. 5 The 'Family Record'—a registry of sayings, reminiscences, etc. by members of a family, in their own voices and of the last words of dying persons. 6 Music boxes and toys. 7 Clocks that should announce in articulate speech the time for going home, going to meals, etc. 8 The preservation of languages, by exact reproduction of the manner of pronouncing. 9 Educational purposes; such as preserving the explanations made by a teacher, so that the pupil can refer to them at any moment, and spelling or other lessons placed upon the phonograph for convenience in committing to memory. 10 Connection with the telephone, so as to make that invention an auxiliary in the transmission of permanent and invaluable records, instead of being the recipient of momentary and fleeting communications. [85]

Edison believed the cost of a telephone would preclude its general use, but he had a genuine concern that his inventions should serve the whole community, not just the affluent. Hence, suggestion ten. The impecunious, by having their messages recorded in their own voice, could then, for a small fee, have them played over the phone.

In England Mr Gladstone opined: 'As to the future consequences it is impossible to anticipate them, all I see is that wonders upon wonders are opening before us'.[87] But Edison's early machines were barely marketable. The Bell Association, in a classic case of the biter bit, made immediate improvements. The tin foil was replaced by a cardboard cylinder, coated with wax, into which the grooves were cut; and Edison's rigid needle was replaced by a loosely mounted stylus. In place of the name phonograph, a new one: the graphophone. By 1887, Bell's innovations had improved Edison's sound quality as much as Edison had improved Bell's in the telephone. It also caused yet another Bell-Edison confrontation as Columbia (Bell's company) set up in opposition to Victor (Edison's, later RCA).

But Edison's sanguine hopes were destined to disappointment. The instrument, marvellous as it was as a triumph of acoustical science, and valuable as it proved to be in demonstrating certain laws, was, beyond these uses, a mere curious toy . . . after so much promise, the performance is so small.[88]

Edison's improved phonograph

The most important improvements came from the third member of the 'who-invented-and-stole-what?' telephone saga: Emile Berliner. In Berliner's opinion:

The new Edison phonograph and the graphophone appear to be practically the same apparatus, differing only in form and motive power.'[89]

By 1888, Berliner had introduced a simpler and more reliable way of cutting grooves, laterally instead of vertically; a method of chemically reproducing a recording from a master take; and the flat, hard shellac disc to supersede the clumsy cylinder. Only then was mass-production for a mass market made easy. And Berliner, at a time when the phonograph was seen chiefly as potential office equipment, saw that it could be used principally for 'home entertainment'.

A talking machine for the whole family at so low a price that it is brought within the purchasing power of everybody.[1]

In 1894, the first disc records were put on the market: 'The Old Folks at Home', 'Marching Through Georgia'. The market boomed.

High School stenography and typewriting, 1904. Remington typewriters and Edison recording machines. Girls outnumber boys in the class by two to one.

A machine so simple that even a child can make it pour forth the most enchanting selections of the world's greatest musicians, singers, actors and speakers.[90]

The novelty ceased to be one, and the instrument found another new name: The gram-o-phone. According to Edison's first biographer, W. K. L. Dickson:

Its many-sided usefulness, such as few inventions are able to show, entitles it to a place in future ages such as it is difficult to conceive will ever suffer displacement at the hands of progressive science.[85]

That is, until the cassette recorder.

When you are experimenting and come across something you don't thoroughly understand, don't let it rest until you run it down. It may be the very thing you are looking for or it may be something more important.[86]

Edison's ability to wring every possible permutation from one single invention was remarkable. One of the more fanciful spin-offs from the phonograph was the phono-motor, a device whereby a sewing-machine, for example, could be operated by sound instead of foot treadle; a diaphragm transferring the sound vibrations of the human voice to a flywheel. The more garrulous the woman, the faster she sewed. 'Thus, if the motor is transferred to a mechanical saw, the tirades of a virago may ultimately find expression in a pile of neatly sawn logs'.[85]

Less whimsical was the electric pen, evolved from the telegraph.

If the intelligent public would imagine the 'business end' of an irate wasp, suspended between the operator's thumb and finger, and venting its pent-up feelings by stinging a succession of holes through a sheet of paper, some idea might be gained . . . of its workings.[85]

The pen's vibrating needle wrote directly on to a sheet of paper which then was used as a stencil from which it was possible to take three to five thousand copies. Whilst Remington was still trying to find a market for the typewriter, 60,000 electric pens were sold to offices in two years. 'Laziness', said Edison, 'is the Mother of Invention.'

In 1878, Edison began work on the electric light bulb. He ended up inventing an entire, commercially viable domestic electricity system. Electric lighting was not new. In Britain, Humphrey Davy had shown in 1808 how a brilliant white light (the arc light) could be produced by an electric current flowing between two carbonised rods. But the voltage needed was enormous. Arc lighting only became a practical proposition for illuminating streets when a dynamo was invented that could generate sufficient,

Street lighting by natural gas, Findlay, Ohio, c. 1880

steady direct current. Lights went on in the late 1860s and early 1870s in London's Trafalgar Square, the Place de la Concorde in Paris, and in the streets of New York, Cleveland and other US cities. But the majority of streets, if they were lit at all, were lit by gas. So were the interiors of houses. Arc lights were too bright and only worked efficiently on a large scale. If there was to be a solution, it seemed to lie in some method of dividing the light and channelling it into smaller units, as was done with gas. Many attempts were made, and all failed. A British government committee reported in 1878:

The sub-division of the electric light is a problem that cannot be solved by the human brain.

The Edison Electric Light Company was formed the same year. From the outset, it had a very clearly defined commercial aim: to capture the lighting market by challenging the supremacy of dirty, smelly, dangerous gas.

As the result of the work of previous experimenters on the problem of subdivision, Edison was left with the choice of concentrating either on the arc light or the 'glow lamp'. Glow lamps, which produced an incandescent light when electricity caused a filament to glow white-hot, offered the most likely solution—though none had ever burnt for more than a few minutes. Edison needed a globe with a good vacuum so that the filament wouldn't flame in the air, and a filament that would offer a high resistance to the electricity. Early attempts with a filament of platinum wire produced minimal light and a quick melt every time he increased the voltage. He turned to the idea of coating a filament with carbon, as this would help resist the heat of the increased voltage. He tried carbonised paper—it lasted eight minutes. Next carbonised wood, fishing line, leather, violin string.

Somewhere in God's Almighty workshop there is a dense, woody growth with fibres almost geometrically parallel and with practically no pith, from which an excellent strand can be cut.[85]

Sewing cotton? On 19 October 1879, Edison tried carbonised cotton threads, bent in the shape of a horse-shoe. They snapped. On the 21st, at the ninth attempt, a cotton filament was finally clamped inside a globe. The result was dramatic. It burned 'without any apparent waste of carbon'. The voltage was increased. It still burned.

It burned like an evening star for forty-five hours, then the light went out with an appalling, unexpected suddenness.

123

Patent No. 223,898 registered the official birth of the modern electric light bulb.

In December, Edison tried again, with his original material, carbonised paper, and with an improved vacuum in the globe. The lamp burnt for 170 hours. The *New York Herald* made it the lead story on Sunday 21 December 1879.

EDISON'S LIGHT

The Great Inventor's Triumph in Electrical Illumination

A scrap of paper

It makes light without Gas or Flame, Cheaper than Oil

Success in a Cotton Thread.

The New York Times, a week later, carried a lengthy interview with Mr Edison, 'a short, thick set man with grimy hands' who showed 'no lack of enthusiasm or confidence'.

Said Mr Edison, 'As there is no oxygen to burn you can readily see that this piece of carbon will last an ordinary life-time. It has the property of resisting the heat of the current of electricity while at the same time it becomes incandescent and gives out one of the most brilliant lights which the world has ever seen. . . . You can see that the electric light is perfected and that all the problems which have been puzzling me for the last eighteen months have been solved.'

In the same newspaper, Professor Henry Morton protested against

the trumpeting of the result of Edison's experiments in electric lighting as a 'Wonderful Success' when everyone acquainted with the subject will recognise it as a 'Conspicuous Failure'. It has been found utterly impossible to render lamps reliably permanent.

Edison's first light bulb

Edison, with an eye on the commercial future, replied to his detractors by inviting the public to a 'Grand Illumination' at his laboratory at Menlo Park. Three thousand people 'impelled by commercial and scientific interest' were brought by special trains. They arrived to find the place in darkness. Then the lights were switched on—seven hundred of them, powered by dynamo. The press called it 'A Heavenly Spectacle', and gas shares tumbled on the New York and London Stock Exchanges. Shares of the Edison Electric Light Company rose correspondingly, from $100 each to $3000.[85]

Edison was still not satisfied. He continued his search for the perfect carbon filament, 'a filament of such inordinate resisting power as to secure a *perfect* sub-division of electric light'. Human hair? Bamboo? Bamboo!

W. H. Moore Esq. Rahway N.J.

Dear Sir,

Your trips to China and Japan on my account, to hunt for bamboo or other fibres, were highly satisfactory to me, as evidenced by the fact that you found exactly what I required for use in connection with my manufacture of my lamps. . . .

Very truly yours,
Thomas A. Edison.[85]

The jungles of the world were ransacked for a satisfactory supply —again, with maximum publicity. Mr Frank McGowan, 'a gentleman of Celtic extraction . . .', pursued the course of the Amazon for 2300 miles, tasted no meat for 116 days and did not change his clothes for ninety-eight. The *New York Literary Journal* declared:

No hero of mythology or fable ever dared such dragons to rescue some captive goddess as did this dauntless champion of civilisation.

After discharging his commission, Edison's 'irresistible lieutenant' was reported missing—'dead', said some; eloped with a native maiden, said others. Either way, the Japanese bamboo continued to serve as filaments for eight years.

But the electric lamp was just a small part of a major scheme.

My idea is to have central stations to cover, say, a square of three or four blocks . . . our electricity will go from our stations just as gas flows from the meter. Whether the company will charge for the light according to the amount of electricity which each consumer uses or whether so much a month will be charged will be determined when the electricity is introduced.

Before the idea could be realised, Edison had to invent switches, sockets for the light bulbs, fuse boxes to make the power safe, meters and underground cables such as Bell had envisaged using for the telephone, and 'a thousand details the world never hears of'.[1] Perhaps most important, there had to be new dynamos to cope with the constantly changing demand for current. General opinion was that such dynamos couldn't be made. Edison showed that they could.

In 1881 came the first urban area to be lit by incandescent lamp: Holborn Viaduct, London. Two Edison generators and nearly two thousand lights, some for the Head Post Office at Mount Pleasant. The following year, in September, a much bigger scheme centred on Pearl Street, New York. Edison and his company had spent $600,000 on the venture, 'the biggest and most responsible thing I had ever undertaken'.[86] Edison started the generators himself.

When the circus commenced the gang that was standing around ran out precipitately and I guess some of them kept running for a block or two . . . there was no parallel in the world . . . all our apparatus,

New industry from the new invention— lights for house and street, electricity meters, a generating machine—all developed at Menlo Park, 1880

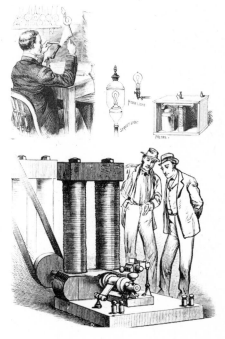

Laying power cables in New York streets, 1882. Electric wires, to provide domestic lighting, were fed through tubes. Joints were soldered and tarred and then tested for insulation.

devices, and parts were home devised and home made. Our men were completely new and without central station experience. What might happen on turning a big current into the conductors under the streets of New York, no-one could say.[86]

The noise and flame was as though the 'Gates of Hell had opened up'—but there were no problems. The lights went on. Not only in the homes of the eighty-five subscribers, but, symbolically, all over the world. It was the start of the Age of Electricity. Soon, electricity would not only light streets, but heat houses, drive railroads and power machines—become, in fact, the world's chief source of energy. And Edison was just thirty-five.

Edison brought to the problem of invention a system which has its parallels in mass-production. His research laboratory was not a boffin's retreat but a factory with the methods and aims of a factory: to produce new goods cheaply for a market. His factory was a place of invention. Not discovery. He made a clear distinction between the two:

Discovery is not invention. A discovery is more or less in the nature of an accident.

Edison discovered the phonograph, he invented everything else, taking other people's ideas and developing them for profit. From 1876, first at Menlo Park and later at Orange, New Jersey, he 'invented to order'. There was no room for 'accidents'. Accidental discoveries couldn't be programmed and time-tabled; inventions could.

Just as the American System divided manufacture into separate stages, so Edison divided his laboratories into separate shops, each shop researching and developing just one specific area of a total problem. Edison's machines were the members of his research teams—hundreds operating according to his personal direction.

The immediate advantage was that research could be thorough: no less than 600 fibres were tested for the electric lamp. And quick: the invention process was radically speeded up; the time from conception to application in the case of the light bulb was a mere three years. And the setup could be a source of job satisfaction:

Edison made your work interesting. He made me feel I was making something with him. I wasn't just a workman. And then, in those days, we all hoped to get rich with him.[86]

Edison was also singular for his enormous vision. He saw the whole before he saw the parts. Bell conceived the idea of a telephone network *after* he'd invented the telephone, as a logical

Marinelli office, New York, 1908—the typewriter, the telephone and now the electric light combine to form the basis of modern office life

extension of his original idea. Edison envisaged an electricity network *before* he'd invented the light bulb. The network itself was the idea, the light bulb and the dynamo were the means of realising it. It marked a radical difference of approach. Bell was applying his invention; Edison worked out the application before he'd started to solve the problems. It was the essential concomitant of a commercial enterprise—and Edison had learnt his commercial lesson the hard way. His first-ever patent was for a vote recorder. It worked well: it was the obvious answer to all the time-wasting manual voting systems in every political assembly. But the assemblymen preferred to waste time. There was no market.

Ever after, I investigated minutely the necessity of any particular invention, before I attempted its reduction to practise. To this decision I have made it the rule of my life to adhere.[85]

That electricity made its domestic appearance in Pearl Street was no accident. Edison had conducted extensive market research. Every house in the area was visited, every opinion canvassed, all the existing gas equipment checked and measured. Only when the scheme seemed commercially viable did Edison act. In a reversal of the traditional situation, no longer was an invention in search of a market—the sewing machine, the typewriter, etc.— but a market was found, and nurtured, and made ready for an invention.

Having entered the commercial jungle, Edison used jungle tactics. Promotion was clever. In New York, there were torchlight processions, each marcher with an electric bulb on his head. At Daly's Theatre, batteries were concealed in dancers' bosoms, and powered 'Glow-worm lights' at the tips of their 'scintillating fairy wands' to give a 'vision of haunting loveliness'.[85] In London, at the Savoy Theatre, each fairy in *Iolanthe* 'carried a small accumulator on her back half-concealed by her wings, and this gave electricity to a miniature glow-worm ornament on her forehead . . . and Sir Arthur Sullivan, during a part of the performance when absolute darkness was necessary, enabled the performers to follow his beat by having the top of his conductor's baton furnished with a little incandescent lamp'.[88]

Competition was ruthless. One of the victims was George Westinghouse. After his success with the air-brake and railroads, Westinghouse pioneered gas lighting in the streets of Pittsburg. He had invented much of the equipment—the gas meters and leak-proof pipes—and these had been adopted in most of America's major cities. In the late 1880s, when the outlook for gas was bleak, Westinghouse switched to electricity and pioneered the use of alternating current (to achieve more power at higher voltage over longer distances). At the same time, the State of New York was contemplating abolishing the 'barbaric punishment of death by hanging in favor of a more humane and scientific method of execution'. The State Commissioners turned to Edison for help.

Edison was against capital punishment, but accepted it as a

Execution by Electricity—the chair and apparatus used in producing death—1889. But old methods die hard—lynching at Fossil Creek Bridge, Russell, Kansas, 14 January 1894.

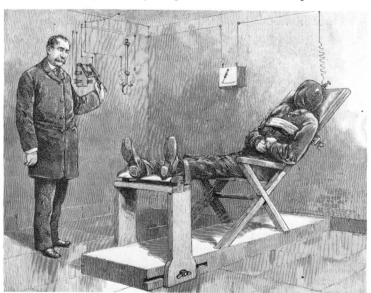

necessary evil. He carried out experiments. The victims were several dogs, a horse and some calves and, by accident, very nearly a member of the research team. 'The results were instantaneous; the slightest possible twitching of the muscles was apparent, but beyond this there was no cry or anything to indicate suffering.' The public approved:

The criminal will be seated in a chair and the mere touch of a button will turn him into a corpse.[88]

From 1 January 1889, the calcined cockroaches of Edison's youth would be succeeded by the calcined bodies of human beings.

Edison, finding his own direct current unsatisfactory for the job of execution, advised the Commissioners to use alternating current. So the State used alternating current. The Edison faction, commercially committed to direct current, were now happily able to equate alternating current, the rival Westinghouse current, with instant death. And instant commercial death for Westinghouse. Westinghouse fought back:

The alternating current will kill people, of course. So will gunpowder, and dynamite and whisky . . . but we have a system whereby the deadly electricity of the alternating current can do no harm unless a man is fool enough to swallow a whole dynamo.[60]

We are still partly saddled with the fallout from the Current War: one part of the world on a/c and the rest on d/c, with electrical appliances having to be adapted accordingly.

Despite the commercial free-for-all, Edison still saw the role of the inventor as the benefactor of society. 'My first impulse on taking any apparatus into my hand . . . is to seek a way of improving it.' It justified his plagiarism. 'Everybody steals in commerce and industry. I've stolen a lot myself. But I know *how* to steal.'[85] And *why* to steal. The environment may no longer be hostile, but it could still be improved:

We will make electric light so cheap that only the rich will be able to burn candles.[16]

Bulbs sold for 40 cents. The cost of production was $1.25. In three years it fell to 22 cents. The bulb still sold for 40. Society and Edison were both happy.

Edison's concept of the research laboratory was as significant for the future as any of the individual inventions that came out of it. It set the standard for the commercial laboratories soon to be set up by all the major American manufacturing companies. The idea of invention as the activity of a lone genius was dead.

George Eastman—the inscription, in East-man's own handwriting, reads 'Made on paper with a soluble sub-stratum developed after transferring. Feb 1884'

A section of Photographic Hall at the Centennial Exhibition, 1876. Photographs were still meant to look like paintings—family portraits or scenic views.

Edison ranged over whole areas of technology—George Eastman (1854–1932) concentrated on one, and helped photography develop from the glass plate to the movies. The same Humphrey Davy who had demonstrated the arc light in 1808 had, six years before, shown how the shadow of an object thrown onto a piece of chemically treated paper remained after the light source was removed. In the late 1830s, Frenchman Louis Daguerre discovered how to prevent the 'shadow' from fading. The Englishman Henry Fox Talbot discovered how to make more than one positive photograph from one negative. By 1860, 'portrait parlors' in the United States were very popular, and Matthew Brady, in the American Civil War, showed how a photograph could capture the harsh realities of a subject in a way that was quite impossible with paint.

But photography was not a field for the amateur. The aesthetics of composition were subordinate to the mechanics. The photographer needed the dedication of the professional, the knowledge of a chemist, and the constitution of a weight-lifter. Equipment was so complicated and so heavy (average weight 100 pounds) that it had to be transported by wheelbarrow or, as in Brady's case, a horse-wagon. The problem was the wet-plate process. As a preliminary, a glass plate had to be painted with collodion (gun cotton in a mixture of alcohol and ether) to form a thin, transparent film, and then be made sensitive to light by being bathed in silver nitrate. The photograph had to be taken with a large tripod camera and developed before the plate dried—all from within the heat and fastness of a black tent. The process was not only cumbersome, but expensive. Brady was pauperised. Yet interest in photography continued to grow.

The Photographic Hall at the Centennial drew vast crowds. There were nearly 3000 groups of photographs from sixteen countries. Displayed alongside was the latest equipment—mostly wet-plate improvements, but also a dry-plate. This was a British invention. Instead of collodion, plates were coated with gelatin and bromide, and could be developed at leisure. It put an end to the wheelbarrow.

Eastman, at twenty-three, had the fanatical dedication to photography of a man whose hobby is a satisfaction substitute for his job. He was a bank clerk.

Being an amateur was, I suppose, arduous work, but one never finds a hobby hard-riding and I went out taking photographs whenever I could, read everything that was written on the subject, and generally tried to put myself on the plane of the professional photographer without, however, any idea of going into the business of photography.

Since I took my views mostly outdoors . . . the bulk of the paraphernalia worried me. It seemed that one ought to be able to carry less than a pack horse load.[94]

Eastman paused for reflection.

At first I wanted to make photography simpler merely for my own convenience but soon I thought of the possibilities of commercial production.[1]

It was the natural thought for any go-ahead young American, and it led Eastman to improve the dry-plate process, devise a machine for coating the gelatine plates, and set up in business. Europe had done the spade-work; America, as usual, supplied the enterprise and proceeded to build the edifice.

> And soon or late I always call
> For stock-exchange quotation—
> No scheme's too great and none too small
> For companyfication.

Gilbert and Sullivan, in *Utopia Ltd*, mocked the American trend. A trend that had galvanised into an obligation.

Eastman 'manufactured on a large scale . . . to put the price down'.[94] It was the American System of Manufacture applied for the first time on a large scale to a non-mechanical product and, by 1884, Eastman was a success. But, like Edison, he continued to make improvements. The glass plates were heavy: he looked for an alternative and found it in strips of paper that he covered first with collodion, to produce a film, and then coated again with photographic emulsion. He wound the paper strip onto a roller that could be fitted into a camera. Surprisingly, it was the roller as much as the film itself that was to play the significant part in future developments.

Reaction to Eastman's improvements in the professional market was muted. He was forced to create another market—an amateur one—or go broke. But the amateur had no skill. To accommodate him, Eastman produced the Kodak: 'an instrument which altogether removes from the practice of photography the necessity for exceptional facilities or, in fact, any special knowledge of the art.'[94] As the Kodak booklet explained:

Today, photography has been reduced to a cycle of three operations. 1 Pull the string. 2 Turn the key. 3 Press the Button. This is the essence of photography and the greatest improvement of all: for where practise of the art was formerly confined to those who could give study and time and room, it is now feasible for everybody. The Kodak Camera renders possible the Kodak System.

'You Press the Button and We Do the Rest'—Kodak's complete processing and developing service

Fixed focus, single shutter speed, and no view finder. Eastman then proceeded to give an object lesson in how to produce and sell a product that people barely realised they might want. They could not be said to 'need' it, in the same way as reapers, telephones or electric lights. It was a toy, a luxury item—one of the mechanical first. Fortunately for Eastman, Americans had been brought up on a diet of novelty, and without a constant supply would have suffered withdrawal symptoms. The novelty Kodak was an instant success and the hard-sell slogan, 'You Press the Button and We Do the Rest', one of the most successful commercials in advertising history. It even got into *Utopia Ltd*:

> To diagnose
> Our modest pose
> The Kodaks do their best;
> If evidence you would possess
> You only need a button press
> And we do all the rest.

What Eastman offered for $25 was not just a camera, but one hundred exposures and—another innovation—after-sales service. For no extra charge, the camera was returned to the Kodak factory for the film to be processed, and for just another $10 sent back to the customer already loaded with another film.

'If it refuses to work . . . it will be repaired free of charge.'[94] Having created his market, Eastman made sure he kept it. It was a system for self-perpetuating profits. Later, as people became more photographically adventurous, the system and the commercial were modified: '. . . We Do the Rest, or you can do it Yourself.'[1] And the do-it-yourself equipment was, of course, manufactured by Eastman.

Like so many of the other big, ruthless American manufacturer/inventors, Eastman was preoccupied with the notion of doing what he did for the benefit of society. With some of those others this was mere verbal justification for their methods and their money, but with Eastman it was genuine. He practised what he preached. He was a great philanthropist and a great humanitarian: a supporter of education and of preventive medicine, he donated heavily and anonymously to institutes and hospitals. Most of all, he was aware of the importance and the social implications of the new leisure.

What you do in your working hours determines what you have: what you do in your leisure time determines what you are.[94]

A simple equation explained and justified it all:
More mechanisation=more leisure
More mass-production=more money to enjoy more leisure.
The camera 'enriched the leisure hours of many'. The luxury item had a new social function.

In 1889, Eastman replaced his paper-backed film with celluloid, which was more flexible, less easily damaged. Celluloid was patented by John and Isiah Hyatt in 1870; its use as film was pioneered by an obscure Episcopalian clergyman from New Jersey and, at the same time, Eastman's own research lab. It was to have a profound influence on Thomas Edison:

In the year 1887, the idea occurred to me that it was possible to devise an instrument which should do for the eye what the phonograph does for the ear, and that by a combination of the two, all motion and sound could be recorded and reproduced simultaneously. This idea, the germ of which came from the little toy called the Zoetrope, and the work of Muybridge and others, has now been accomplished, so that every change of facial expression can be recorded and reproduced life size. The kinetoscope is only a small model, illustrating the present stage of progress, but with each succeeding month new possibilities are brought into view.[85]

The Zoetrope, or 'Wheel of Life', was British.

This rude prototype contains a cylinder ten inches in width and open at the top, around the lower half of whose interior a series of pictures is placed representing a sequence of motion it may be desired to portray, such for instance as wrestling, jumping, or the swift progress of animals. These movements are seen through the narrow vertical slits in the cylinder during the rapid revolution of the little machine, and are designed to blend into one continuous impression.[85]

The No. 1 Kodak camera, 1888

Kodak's first folding camera was introduced in 1890. It could produce 48 four- by five-inch pictures. Advertisements for Kodak cameras now read 'seven styles and sizes'.

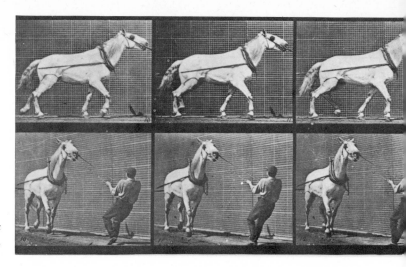

Eadweard Muybridge: A man pulling a horse—from two different camera angles, c. 1887

Muybridge was an Americanised Englishman who had devised a system of taking a sequence of photographs of a galloping horse. His equipment consisted of racetrack, a horse, and trip-wires for the horse to trigger a bank of twenty-four cameras at split-second intervals. The results showed that instantaneous photography was possible. What was not yet possible was to take pictures at sufficiently short intervals in order to get 'the absolute blending of outline essential to a faithful portrayal of life'. Edison got to work, and came up with the Kinetograph, a high-speed camera, US Patent No. 403–534, 1891, '. . . taking a large number of photographs of a moving object in such a manner that any two successive pictures are almost identical'.[1]

Split-second action—800 different images on the fifty-foot roll. Muybridge's photographs were all taken from a slightly different point of view; Edison's all from the same viewpoint. Muybridge used wet-plate; Edison used celluloid. But the use of celluloid and the idea of film on a roller were both Eastman's contributions. Without them, Edison's camera would not have been possible.

In 1893, Edison went into the film-making business. Operations were conducted in the Kinetographic Theatre, the first motion-picture studio, otherwise known as the 'Black Maria':[85] a shed, painted black, with a roof that opened and could be turned towards the sun. A handful of actors, a handful of props, and the Kinetograph. To view the results: the Kinetoscope,

. . . by means of which a single composite picture is seen by the eye, said picture giving the impression that the object photographed is in actual and natural motion.[1]

The Kinetoscope was invented by Edison with the help of his assistant, W. K. L. Dickson. It was simply a wooden box in which the fifty-foot band of film was run continuously over a series of battery-driven rollers, between an electric light bulb and a rotating shutter. The performance was viewed, at the rate

of forty-six frames a second, through a small glass slot fitted with a magnifying glass. Edison wrote to Muybridge:

I am very doubtful if there is any commercial feature in it . . . these Zoetropic devices are of too sentimental a character to get the public to invest in.[86]

He was wrong. The 'chronophotographs of Buffalo Bill, Eastern knife throwers, Kings and Queens of the tightrope and trapeze', were the mainstay of the kinetoscope peep-shows in penny arcades, the 'well-known nickel in the slot'.[85] Viewing time was forty-five seconds. Very successful. But for projecting film on to a screen, the Kinetoscope was impractical. Edison resorted to the familiar ploy of 'improving' somebody else's projector.

In 1895, he demonstrated his method for combining the Kinetoscope with the phonograph to produce films that synchronised sound with picture.

Nothing more vivid or more natural could be imagined than those breathing, audible forms with their tricks of familiar gesture and speech. The organ grinder's monkey jumps upon his shoulders to the accompaniment of a strain from 'Norma'. The rich tones of the tenor or soprano are heard, set in their appropriate dramatic action; the blacksmith is seen swinging his ponderous hammer, exactly as in life and the clang of the anvil keeps pace with his symmetrical movements.[85]

Although his method was out of date by the 1920s, Edison produced the first 'talkies' thirty years before Al Jolson appeared in *The Jazz Singer*.

Edison seemed to be bent on doing for the motion picture business what he had done for electricity—inventing the entire system and its details, like sprocket holes in the film, the size of the film (35mm), even down to casting the actors. But the actual celluloid continued to be the product of George Eastman, specially produced to Edison's requirements. Credit for the in-

Edison's patent for his Kinetoscope, showing the continuous band of film and the interior of his 'Black Maria', the Kinetographic Theatre, Orange, New Jersey. The phonograph (left) is wired up to the the kinetograph (right).

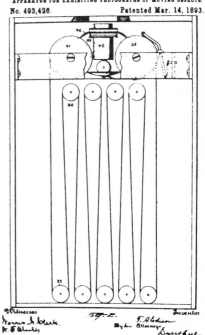

Lyman Howe's 'New Marvels' were advertised in 1898 as an improvement on live-theatre entertainment—a kind of 3D and Cinerama all in one. The real marvel was that Howe was allowed to peddle such blatant misinformation. An audience viewing such an entertainment, if such had existed, would have suffered severe eye-strain and the people in the boxes would have been able to see practically nothing at all.

vention of a new industry—the Movies—must be shared equally between them.

Although the first public film demonstrations were given in France, by the Lumière brothers in 1895 and the following year by George Mélies, the movies became a particularly American form of entertainment. By the 1950s, they had an estimated audience of eighty million. Unlike the middle-class opera and theatre imported from Europe, they were democratic, a mass product that everyone could afford. And everyone could understand—just as the people of the Middle Ages could understand the images in the stained-glass windows of the church. There was no need to be 'educated'.

But the movies were more than just commercial entertainment. They have been described as 'the most potent force for the dissemination of ideas since the invention of the printing press'.[1] They gave America a knowledge and an image of herself, and helped reinforce her ideals and aspirations—as well as perpetuate some of her myths—even to a point of falsification.

'White man good; red-skin bad,' was Hollywood's version of Manifest Destiny. Edison, himself a cowboy of commerce, loved 'westerns'—he obviously associated.

'Poor boy makes good' was another maxim: the movie-moguls

themselves could testify to the continuing truth of that.

America was still the 'Land of Opportunity' where 'any guy can make it to President'. But the glamour began to cloak the reality. There was little mention of the strikes, the unrest, the poverty and the political corruption that was too often the obverse of scientific achievement and the new, good, consumer-product life. Hollywood became the custodian of the American Dream, proceeded to subdivide it like electricity, and peddled little dreams in the form of movies because 'that's what the people wanted'. Dreams, ideals, aspirations—part of the collective national inheritance, but served up long after some of them had begun to wear out.

Movies at Crystal Hall, New York, 1914— the cinema industry already under way. Terms soon to be familiar include 'big feature' and 'terrific entertainment'

Chapter 5

Pioneer Run at Oil Creek, Pennsylvania, 1865. Speculators swarmed all over the neighbouring hillsides, '. . . hundreds and hundreds of them. Everything muddy and dirty . . . the whole smelling like a camp of soldiers when they have diarrhoea.'

The New Power

The Corliss steam engine was the biggest ever built: 700 tons, 40 feet high, 2500 horsepower. At the Centennial it was 'the Marvel of all Spectators'. It supplied power to all the other machines in Machinery Hall. It was the symbol of that 'Great God, that Almighty Power' that drove the factories, the railroads, the boats and the reapers of America's prosperity. Without steam 'the great wheels would remain silent, the busy hum would cease, the machinery would lose life'. Steam had given America control over the hostile land and over her own destiny; given her the power to function and the power to expand. Steam was America's future.

But, tucked away in a corner of Machinery Hall was an exhibit that presaged a power revolution greater than anyone could have imagined. It was a small, working model of an oil well. Oil, not steam, was to shape the American economy, shape American society, for the next one hundred years.

The first oil well was successfully sunk in north-western Pennsylvania in 1859. But oil as a substance was already well known: it had been ruining the salt wells for years and, during the forties and fifties was collected, bottled and sold 'in small quantities at a high price and has entered into the composition of several popular lotions for rheumatism etc.' The 'etc' included cures for anything from bladder trouble to toothache. The patient smeared it on his skin or, reassured by names like 'Rock Oil' or 'Seneca Oil', drank it.

The Pennsylvania Rock Oil Company was formed in 1854. An analysis

The great Corliss Steam Engine—symbol of an age: hope for the future, Centennial Exhibition 1876

The Empire Transportation Company's exhibit at the Centennial Exhibition 1876—a model of oil wells. Oil was first discovered in Pennsylvania in 1859.

by a Yale Professor had recognised oil's suitability as a lubricant and—what no one had expected—as a fuel for use in lamps. Most homes in the early part of the century were still lit by candle, or by the 1830s with lamps burning whale oil. But whale oil was becoming expensive as the whales became scarcer. As alternatives there were various animal and vegetable oils, 'coal oil' distilled from tar, and gas. But gas, expensive to pipe, was the status symbol of the urban rich. Oil—petroleum offered a suitable, less smelly, less smoky but just as efficient substitute for everybody. The problem was how to get hold of it.

The Seneca Indians had provided one answer. They threw a blanket on top of an oily pond till all the oil was soaked up. The blanket was removed and the oil squeezed out. It was hardly a method suitable for commercial production. In 1857, the Pennsylvania Rock Oil Company called in Edwin L. Drake, ex train conductor and steamboat clerk, and now a bogus colonel—'Colonel' to lend respectability and generate confidence in the company's shareholders. Drake had no practical experience of anything remotely technological; nevertheless . . .

. . . within ten minutes of my arrival . . . I had made up my mind that [oil] can be obtained in large quantities by boring as for salt water. I also determined that I should be the one to do it.

He was in the great American tradition of a man able to turn his hand to anything. Operations finally got under way in June 1859, at Titusville on Oil Creek, with the aid of a blacksmith and salt-well borer, 'Uncle Billy' Smith. But the first exploratory holes were flooded with water. To exclude the water, a pipe was driven from the surface right down into the bedrock, and then a drill was fed through the pipe. It had never before been attempted, but it worked, and became the standard technique, still in use today.

On 27 August, the pipe had reached a depth of $69\frac{1}{2}$ feet. Operations were suspended for the weekend: the Sabbath was still sacred. On the Monday morning the pipe was full. Now the legend:

COLONEL DRAKE What is it?
'UNCLE BILLY' It's your fortune coming up.[1]

It was America's fortune, it was a lot of other people's fortunes—ironically, it was never Colonel Drake's. He became a broker in oil stocks on the New York Exchange and lost everything.

To extract oil in sufficient quantities, Drake built a derrick and installed a pump—devices already in use for drawing brine from salt wells—then stored the oil in any vessel he could find. By

October, the well was producing twenty-two barrels of oil a day—
with net profits estimated at over $20,000 a month[101]—till Drake,
peering into the well with a lamp, blew it up.

Wells are sinking in every direction and strangers are flocking in from
all parts of the country . . . every son of Pennsylvania should rejoice in
the good Providence that has enriched the state . . . with rivers of oil.[101]

Oil towns mushroomed overnight:

. . . adventurers flocking in to this promised land . . . thousands of
persons on foot, on horse-back and in sleighs.[101]

Pithole Creek, near Titusville, grew in eighteen months from four
log-cabin farmhouses to a town of 15,000 people.

Drilling, pumping, buying, selling, building . . . everything hurried . . .
all excitement, life and activity.[1]

The hillsides covered with oil derricks, wooden shacks and
storage tanks.

Hundreds and hundreds of them. Everything muddy and dirty . . .
the whole smelling like a camp of soldiers when they have diarrhoea.[100]

A daily paper, two churches, two banks, two telegraph offices and
fifty-seven hotels.

A gigantic city of shreds and patches . . .[100]
. . . the Mecca of the Oil World or the 8th Wonder, I don't know
which . . .[102]
. . . crammed full, two in a bed everywhere.[1]
. . . buildings go up and have groceries in them in six hours . . .[102]
. . . tough beef, bread and a decoction of unknown ingredients called
coffee.[101]

The Astor Hotel went up in a day and a half. Murphy's Theatre
had seats for a thousand, and Tiffany chandeliers.

*The Drake Well at Titusville, Penn-
sylvania, c. 1866. Colonel Drake is the man
in the top hat.*

A bar is the almost invariable appendage to every building. There is more vile liquor drunk in this town than in any other of its size in the world.[102]

Everyone on the make. Lawyers:

There's 'Ketchum and Cheatum' and 'Lureum and Beatem',
And 'Swindle um' all in a row;
Then 'Coax um and Lead um' and 'Leech um and Bleed um',
And 'Guzzle um Sink um and Co.'[102]

Doctors:

Teeth extracted safely, each 50 cents.
Teeth extracted without pain, 1 dollar.[102]

And prostitutes.

Steam-power is rapidly being introduced . . . engines of 5 h.p.[1]

Steam, in effect, drilling its own grave. Peak production, 6000 barrels of oil a day.

> Half an inch, half an inch
>   Half an inch downward,
> Down to the Gates of Hell
>   Bored the Six Hundred.
>
> Petroleum to the right of them,
>   Rock oil to the left of them,
> Coal oil beneath them,
>   Quietly slumbered.

Entitled 'The Successful Well', this lithograph shows the order that should have been. A steam engine driving the well; the oil being pumped into an open vat; the vat discharging oil into a barge to be floated downstream. Good management, no mud, the minimum environmental pollution and time to read a newspaper.

Waved they their leases there
   With a triumphant air,
Each greasy millionaire
   Counted his profits while
All his friends wondered.
   Plunged in the dirty soil
Straight through the rocks they toil,
   While the poor sceptic
Struck by the smell of oil
   Thought he had blundered—
Sad he returned, but not
   Yet the six hundred.[103]

It was a 'second California in north-western Pennsylvania', the Gold Rush all over again. 'It's not uncommon', reported a local paper, 'for a million of dollars to change hands in a single week'.[1] America at last was beginning to tap its vast mineral resources— Pennsylvania now, West Virginia, Ohio, Indiana, California and Texas to come. But at Pithole, the bonanza dried. Within five years, the town was dead. The people picked up their bricks and boards and moved on. Today it is as if the town had never been.

Crude oil had to be refined: it was distilled. It had to be stored: it was put in whisky barrels. It had to be transported. Transportation was the biggest problem. One early solution was to raise the water level of local streams by damming, load the oil on to boats, break the dam and, for a few precious minutes, turn the stream into an effective canal. Another was to drag the barrels on wagons through the mud. This inconvenience led the industry to produce its second major technological innovation— the pipeline—and to run one (the world's first) from Pithole to a nearby railhead. But in the main, the embryonic oil industry was a technical, organisational and financial shambles. Speculators fought with producers, who in turn fought with the railroads.

Almost everybody you meet has been suddenly enriched or suddenly ruined (perhaps both within a short space of time), or knows of plenty of people who have been.

The price of oil rose and fell as fast as the boom towns that produced it:

DATE	PRICE PER BARREL
1858	$30
1861	$1.35
1864	$14
1870	$2.70

The man who brought order to chaos was John D. Rockefeller, who gave the word 'trust' an entirely new meaning:

'Firing the wonderful thousand-barrel well on the Arbuckle homestead.' The Arbuckle Well, Pittsburgh, Pennsylvania, became a national legend.

Rockefeller, 'mad about money but sane about everything else,'[108] was born in 1839. Twenty years later, in the year of the oil rush, he was in Cleveland, Ohio, dealing in meat and grain and half-way to being a millionaire. In 1863, he bought his first oil refinery and was soon turning out lubricants for machines, and kerosene ('coal oil') for lamps. By 1865 he'd made his first million. But fluctuations in the oil price made his second million difficult to guarantee. As an insurance, he invented his own personal version of Manifest Destiny, wherein money was to be made quite simply by destroying the opposition. Rival businesses became the Indians of Rockefeller's commercial Wild West.

Rockefeller's secret was to cut free from dependence on anyone else. To make his own oil barrels, he bought forests. To make his own chemicals, he bought factories. To ensure transport he bought railroad cars and ships. He drove a deal with the railroads for cheap price for bulk haulage and, because his business was now so important, got the railroads to pay him a percentage of the profits on oil they shipped for his rivals.

Rockefeller was able to name his own price because, by 1872, of twenty-six oil refineries in Cleveland, he owned twenty. As he was the main refiner, he called the tune with the producers—and he never produced himself, it was too risky.

Under the thin guise of assisting the development of oil refining . . . this combination has laid its hands on the throat of the oil traffic with a demand to stand and deliver. [*New York Tribune*]

'Monopoly' was a game invented by Americans and, in the last years of the nineteenth century, the nation played it for real. In 1872, President Grant said:

I have noticed the progress of monopolies and have long been convinced that the national government would have to interfere and protect the people from them.

Rockefeller was not deterred. In 1876:

The coal oil business belongs to us: we have sufficient money laid aside to wipe out any concern that starts in this business.

And, with even more sinister overtones, in 1879:

I have ways of making money you know nothing of.[108]

In 1880, the investigating Hepburn Committee of New York reported that Rockefeller's Standard Oil:

. . . owns and controls the pipe-lines of the producing regions. It ships 95% of all the oil and dictates terms and rebates to the railroads. It overbids in the producing regions and undersells in world markets. It

John D. Rockefeller carried his passion for perfection onto the golf course. He hired a boy to do nothing but shout, on each shot, 'keep your head down'.

buys out and freezes out all the opposition, until it has absorbed and monopolized the entire traffic.

Rockefeller took refuge in the Standard Oil Trust. All stockholders, in return for certificates, secretly handed over their shares to nine trustees. This empowered the trustees to run the company without any 'democratic' brake whatever. To undercut competition the trust reduced its oil prices—'to such an extent when any other oil is offered that they force the parties handling the oil to abandon the trade'.[106] When the smaller outfits went broke, Standard 'put their oil up again and quote at the old price'. If this tactic failed, then the Trust proposed an 'attractive' deal or a takeover, described by the unfortunate business concerned as 'the gentle fanning of the vampire's wings [before] the undisturbed abstraction of the victim's blood'.[108] And if that failed, too, the Trust's industrial espionage network went into action—with threats, arm-twisting (metaphorically and literally), and breaking heads.

The press, one of the keepers of the American conscience, was outraged. *Harper's:* 'It has crushed out competition and honest industry.' Oil City's *Daily Derrick*: 'He is the Mephistopheles of Cleveland.' On stage he was attacked as 'the greatest criminal the world ever produced'.[108] Cartoons showed him as an octopus or a hydra without a heart.

Rockefeller was offended. Like so many other get-up-and-go nineteenth-century American businessmen, his business zeal was paralleled by a religious one. He saw no inconsistency between the two; indeed, one was an extension of the other. A devout Baptist, he often spoke from the pulpit:

Accomplishment. That's the goal of every man who tried to do his

'And he asks for more!'—anti-monopoly caricature from 'Puck', 7 May 1890

part in the world. Put something in and according as you put something in, the greater will be your dividend of salvation.[108]

There was the sincerely felt concern for society:

I believe it is my duty to make money, and still more money, and to use the money I make, for the good of my fellow men, according to the dictates of my conscience . . . and we must ever remember we are refining oil for the poor man and he must have it cheap and good.[108]

The other adherents to the other Manifest Destiny also justified their behaviour by a recourse to God.

The oil monopoly qualified as an un-American activity on two counts. One, its exploitation of a natural resource for private profit was in direct conflict with the American ideal that resources should benefit the entire community. Two, it put an end to a local free enterprise that had flourished for a hundred years. Either way, it was a denial of individual rights—except the loudly protested rights of the fortunate few. For Rockefeller was not alone. There were three hundred Trusts, and 'ethics' and 'morals' were not words for the boardroom. There was the ruthlessness of the 'beef barons' of Chicago; the graft of the politicians of New York and Washington; the extortion of Gidden's barbed-wire monopoly. And the blackmail of the railroads. In practical terms:

They start their railway track and survey their line near a thriving village. They go to the most prominant citizens and say 'If you will not give us so many thousand dollars we will run by.' And in every instance where the subsidy was not granted, this course was taken, and the effect was just as they said, to kill off the little town. [Constitutional Convention, 1878]

In metaphysical:

The political power of the railway corporations . . . is a matter as well known as is the corruption by which it was acquired. . . . They drag their slimy length over our country, and every turn in their progress is marked by a progeny of evils. Thus is our land cursed. [*Harper's*, 1868]

It was the invention of a new kind of American businessman: 'Hardened by competition and impersonal because his successes came from selling the products of machines.'

The people began to fight back. In 1877, in the face of a ten per cent wage cut, railroad employees in the east went on strike. In 1892, the strike at Homestead was put down by a show of federal force. In 1894, Pullman, King of Luxury, announced an eight per cent dividend to shareholders and a twenty-five per cent wage cut for railway men. The strike that followed ended in widespread violence throughout Chicago. The old bonds between worker and boss, that had existed for a century, broke down under the pressure of mechanisation and self-interest. The early American ideal of freedom and fair shares for all seemed suddenly to have polarised into the sectional interests of the Haves and Have Nots. But it was a situation that most Americans were prepared to live with. American labour did not unite and organise itself as it did in Britain and the rest of Europe. The fight was spasmodic. Too many members of the unions and brotherhoods aligned themselves with the aspirations and success of their bosses, a situation that had its roots in the comment of the Edison employee, '. . . in those days we all hoped to get rich with him'. The American Dream had a few years still to run.

In 1890, Senator John Sherman, from Rockefeller's own state of Ohio, master-minded an act that forbade every 'combination in the form of trust or otherwise, or conspiracy in restraint of trade or commerce'. Daniel Drew, forger, robber and controller of the notorious Erie Railroad, complained, 'The trouble with the United States is, it's too democratic'.[1] Rockefeller reacted differently:

Look at that worm. If I step on it, I call attention to it. But if I ignore it, it disappears.

The anti-trust legislation was not very effective. It was easily by-passed, and in any case had come too late to protect the small man —he'd gone to the wall already. In 1920, the major companies controlled over half of the oil-refining industry. By 1960, they'd increased their share to eighty-three per cent.

By the 1880s, the American genius for invention was equalled by her genius for business organisation. It was inevitable. Once the outlet for inventions became mass-production and mass markets, organisation was at a premium. Rockefeller's contribution to American invention was to complete an idea that Lowell and Whitney had begun. As Lowell had put the process of cotton manufacture all under one roof, Rockefeller erected that roof over a whole industry. And the inventor, from now on, was to remain in its shadow.

Rockefeller was synonymous with really *big* 'big business'. But 'none of us ever dreamed of the magnitude of what proved to be the later expansions'. It came about as the result of the experiments of George B. Selden (1846–1922). In 1879, three years before Rockefeller formed his Trust, Selden applied for a patent on 'an internal combustion engine designed to propel a horseless carriage'. But, said Henry Ford thirty years later, 'There's very little new under the sun'. Selden believed he had invented the motor car. He hadn't. Steam-powered horseless carriages were

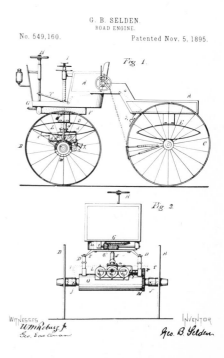

147

operating in Britain in the 1830s. A battery-operated electric carriage was invented there in 1870. In France, a one-cylinder internal combustion engine was operating in 1863. A four-cylinder engine was invented by N. P. Otto in Germany.

George Selden was by profession a lawyer, by inclination an engineer. His legal activities involved helping George Eastman prepare his application for patenting dry plates; his engineering ones aimed 'to work a revolution in locomotion upon common roads'.[32] It was an enterprise in the great American tradition of 'think big'. As Bell had looked back to Morse, so Selden believed his invention would give him 'a place in industrial history analogous to that of the inventor of the steam-engine, the locomotive, the cotton-gin and the telegraph'.[110]

Whilst at the Centennial, displaying his patent machine for barrel-hoop finishing, Selden kept his eyes open for an engine—small, liquid-fuelled but powerful enough to drive the vehicle he had already built in his mind. The gas-driven Otto engine was much too big, nearly ten feet high. Also too big, but more promising, was the Bayton engine made by an Englishman living in Boston. It was the first to use crude petroleum, and could, possibly, be reduced in size. Selden set to work. His objective: to reduce size without reducing power. 1876:

If ever I get a road wagon it will be by accident. Of the almighty effort which an invention requires, who knows but the inventor.[110]

1877:

The road engine comes on slowly. Want of time is the principle difficulty . . . she is going to go, unless I meet some more serious practical problem than I now forsee.[110]

By 1878, Selden had reduced the weight of the Bayton engine by two-thirds—from 1160 pounds to 370—whilst actually increasing the horsepower from a miserable 1.4 to a not quite so miserable 2. The engine had three cylinders and could be fuelled by either gasoline or kerosene (British read petrol and paraffin).

I had thus, as I believed, after a long struggle, demonstrated the possibility and practicability of building a liquid hydrocarbon engine of sufficient power for the purpose of light road locomotion on a common road within the permissible limits of weight and space.[110]

Selden, clever lawyer that he was, applied for his patent in May 1879. It gave him immediate legal protection. He carefully avoided supplying the working model that the Patent Office required as a deliberate tactic to delay the actual issue of the patent itself. Once the patent was granted, it would be valid for only

'We cannot conscientiously feel that Mr Selden ever added anything to the art in which we are engaged.' George Selden in his workshop aged 63.

seventeen years. Selden knew that he was a man before his time. The public was not ready for his engine. What was the use of protection when there was nothing worth protecting? He would wait until someone else began making horseless carriages, then collect his patent—and the royalties.

As Selden waited, the initiative for manufacturing motor-cars passed to Europe. America lagged. On Thanksgiving Day, 1895, the proprietors of the Chicago *Times-Herald* attempted to stimulate American inventors in 'a new and most important field which has been pre-empted abroad and neglected at home'. They sponsored a road race for what were called—for want of a better name —motor-cycles. The race was over a distance of $52\frac{1}{2}$ miles through the main streets of the city. All contestants were issued with road maps and had seven hours in which to complete the course— minimum speed 8 mph. There were eleven entries but only six made it to the start line. These included one of Edison's inventions in the form of the Sturges Electric motor-cycle powered by an electric battery, and the rubber-tyred, tiller-steered, two-cylinder petroleum-propelled car of the Duryea Brothers.

There was $5000 at stake. The race began and continued in a snowstorm. The *Times-Herald* recorded progress:

Many cameras were levelled at the machines as they moved off through the crowds and the slush . . . the Kodaks took snapshots of everything in sight. . . .
The Duryea Motor Vehicle received its first cheer at the corner of Cottage Grove Avenue. Police had to hold the crowd back. . . .
Despite the conditions, vehicles proceeded at a clipping gait into the park where deep snow forced the Da La Vergne vehicle to retire. . . .
The Sturges electricity motor cycle had to make frequent stops to prevent its battery from burning out. . . .
Spectators loudly cheered the Duryea vehicle at North Clark Street but, unfortunately, the operator mistook the hand of the guide post and proceeded at high speed in the wrong direction. . . .
The Macey wagon butted into the rear of a street car and broke its running gear. . . .

. . . at which point only two contestants remained:

The Mueller vehicle was aided by several outside persons in direct contravention of the rules. . . .
The Duryea vehicle called in the help of a blacksmith.

First past the winning post was the Duryea car. Second, one hour and thirty-five minutes later, Mueller. Macey finished on the following day. All should have been disqualified. But first prize went to Duryea, 'for best performance in the road race, for range

J. Frank Duryea at the tiller of the Duryea car, the Chicago 'Times-Herald' race, 1895. Snow made driving difficult.

The first all-American car was Charles Duryea's motorised buggy of 1891. It was tiller-steered, had solid tyres and carriage lamps.

of speed and pull with compactness of design'. Although the race, as a race, was a fiasco, the significance was not in the winning but the travelling. As the *Times-Herald* pointed out, 'The test over a route covered with snow and slush was one to which a horse would never have been subjected'. The race showed conclusively that automobiles were superior to horses and that gasoline was superior to electricity or steam.

Duryea went into immediate commercial production. Within twelve months, the company had produced thirteen cars. The time seemed right for Selden to activate his patent. It was granted on 5 November 1895, sixteen years after the original application: Patent No. 549,160, for an 'improved road engine'. The legal battles began: a rerun of Whitney versus Holmes; Howe versus Singer; Bell versus Edison. They lasted five years. Selden's stand in 1899 was that the increasing number of automobiles—there were, in fact, only fifty in the whole country—were 'exemplifications' of his patent. He spoke of 'his' invention, of his being 'the father of the automobile' and of his modifications of the Bayton engine as 'a substantially new creation'. The judge agreed, despite the fact that the model car that Selden produced to make his point was so front-heavy that it fell over. Duryea, Olds, Winton, Dodge and others came to terms. Selden was paid a royalty on every automobile produced and licences were issued to respectable manufacturers.

Henry Ford's application for a licence was refused. He was a mere 'assembler' of motor vehicles and not a manufacturer at all. Ford, the underdog, the small man, the public's favourite, told the owners of the Selden patent to go to hell and went on assembling.

We cannot conscientiously feel that Mr Selden ever added anything to the art in which we are engaged. We believe the art would have been just as advanced today if Mr Selden had never been born. That he made no discovery and gave none to the world. If he did, it was a narrow and impracticable one having no value, and that he and his assignees cannot monopolize the entire trade by forcing upon it an unwarrantable construction of his claims by those interested in sustaining them.[110]

They were fighting words, prompted by a strong element of that indigenous American belief that discoveries were for community, not individual profit. They appeared in several newspapers and in the new motor magazine, *Horseless Age*, with the added rider that patents merely served to 'exploit the consumer and place a heavy burden on productive industry'. The court cases that followed made newspaper headlines, gave Ford massive publicity, and only ended in 1910 with the judge upholding the patent but ruling that it was no longer applicable. After thirty-three years the automobile had outgrown Selden. He didn't make the fortune or the popular place in history that he had set out to. 'Morally, the victory is mine,' he said. Ford was more to the point:

If we had not won that suit, there would never have been in this country such an automobile industry that exists.

Henry Ford (1863–1947), Assembler of Automobiles. Champion of Mass-Production:

A scientific discovery's a fine thing in itself but it doesn't help the world till it's put on a business basis.

It could be Edison talking—Edison and Ford were great friends. It sums up the new attitude.

Henry Ford, 1863–1947

Ford produced his first 'buggy' in 1892. In his working hours he was employed by the Edison Illuminating Company of Detroit, but in his spare time he was producing another Ford vehicle, the 'autocycle'. It looked like a four-wheeled bicycle with solid tyres and spoked wheels and was driven round Detroit in 1896. In 1903, Ford formed the Ford Motor Company, with the aim of mass-producing a cheap, simple, standardised product. The aim was identical to Singer's, Remington's, Eastman's, Edison's and dozens of other manufacturers'—the main difference, of course, was that Ford's products would be larger and *less* cheap than anything previously on offer. His first product, a two-cylinder, 8 hp car, sold for $850. In 1908, he improved the design and introduced the Tin Lizzie—the Model T. It was so successful that it stayed in production for twenty years.

Ford is important not because he produced motor-cars, but because of *how* he produced them. In his original factory, cars were assembled by half a dozen men fitting the pieces together as they were brought by other workers from different parts of the factory. It was time-consuming and relatively expensive. Ford, an efficiency fanatic, set about becoming an efficiency expert.

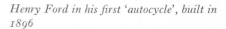

Henry Ford in his first 'autocycle', built in 1896

Like Rockefeller, he owned or controlled the source of all raw materials and their transport. Like Singer and Eastman, he advertised widely. But it is in the actual process of manufacture that, from 1914, he brought together all the elements of the American System of Manufacture, added some ideas of his own, and brought mass-production methods to a peak. Like the mills at Lowell, Ford's new factory at Highland Park, Detroit, operated with a central power source and housed all processes under one roof. Like Whitney, Lowell and Singer, Ford broke down the

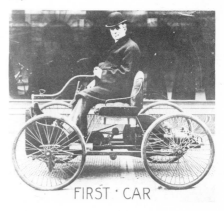

FIRST · CAR

work into separate and simple tasks that could be done easily and quickly. Like Whitney again, Ford believed:

It is essential to economical manufacture that parts be interchangeable. Unless parts fit accurately, the assembly loses motion and much of the economy is lost.

He used standardised parts, both in machines and products. The economy gained was an increase in annual production in the automobile industry from 350,000 before the introduction of interchangeability, to eight million after it. And, as with Whitney's guns, instead of having to buy a new product when anything went wrong, the car owner could just buy a new part.

But Ford's most important technological innovation was the assembly line. The conveyor belt had been used by Oliver Evans in 1804 in his automatic grain-mill—into which he'd earlier introduced the steam engine. But for Ford, 'the idea came in a general way from the overhead trolley that the Chicago packers use in dressing beef'.[112] This was the 'dis-assembly' line that Kipling had seen in 1898. Ford visited the factory of Swift and Co. in Chicago to see the newly automated system at work. The before-and-after automation comparison was dramatic.

Before: 20 men, working an 8-hour day, each killing and completely cleaning one animal before tackling the next. Daily output, 620.

After: 20 men, working an 8-hour day, each man with one specific task as the overhead conveyor belt passed by them. Daily output, 1440.

It convinced Ford. With machines now deployed as effectively as possible, the only inefficient machine was the worker.

Efficiency, by the turn of the century, had become an American bug. Frederick Winslow Taylor, in *Principles of Scientific Management* (1911), became its high priest. He made hundreds of studies, mostly in the steel industry, watching how men moved, the speed at which they moved, how long it took them to do the same job under different conditions with different kinds of tools. He showed how tools and machines could be adapted to get the most work out of a man, and how man should be trained in order to get the most out of the machine:

The correct speed is the speed at which men can work hour after hour, day after day, year in and year out, and remain continuously in health.

The implications were far-reaching:

In the past, the man had been first; in the future, the system must be first.[31]

The model T in action—capable of 45 mph and 20 mpg. By 1914, Ford had one sixth of the nation's automotive labour force, produced one half the nation's cars and made a profit of 80 million dollars.

Man must be subordinate to the machine.

Taylor expounded the principle; Ford put it into practice.

The first step in assembly came when we began taking the work to the men, instead of the men to the work. We now have two general principles in all operations—that a man shall never have to take more than one step, if it can be possibly avoided, and that no man need stoop over.

Chicago's pigs became Detroit's car parts.

Every piece of work in the shop moves; it may move on hooks or on overhead chains, but going to assembly in the exact order in which the parts are required. . . . Place the tools and the men in sequence of the operation, so that each component part shall travel the least possible distance. When the workman completes his operation, he drops the part always in the same place, which place must always be the most convenient to hand. The net result of the application of these principles, is the reduction of the necessity for thought on the part of the worker, and the reduction of his movements to a minimum.[112]

This continuous, flowing line of production was Ford's greatest achievement. As the car bodies at Highland Park slid onto the car chassis, they were bolted together and the finished automobiles driven off, to sales points, under their own power. Assembly time was cut from thirteen hours to one and a half. By 1924, for every two cars in the world, one was a Model T.

Ford was one of the last great entrepreneur-inventors. Soon, American production would pass, already was passing, into the hands of the faceless industrialists who would run the multiple combines. But Ford had great faith in the future:

We have only started on our development of our country—we have not, as yet, with all our talk of wonderful progress, done more than scratch the surface.[113]

His words were a strange echo of those of President Grant in 1876.

Ford was also very aware of America's past. When the rest of the nation was smashing up its history, he, the 'history is more or less bunk' man, was busy collecting it and preserving it in his own private museum at Dearborn—ploughshares, telephone circuits, locomotives, etc. Pullman bought the Corliss in 1911 and broke it up; Ford bought a smaller Corliss and put it in his museum for posterity.

The only history that is worthwhile is the history that we make day by day . . . the history of our people, as written in the things their hands made and used.

Ford assembly line, Highland Park, Detroit, 1913

Ford's first moving assembly line (right) at Highland Park, 1913. On the outside of the factory, autobodies were lowered and bolted onto chassis and the final cars driven away.

Automobiles outside the Vanderbilt mansion at the start of the New York to St Louis Tour, 1904

Smith and Mabley's autoshowrooms, 1765 Broadway, New York, in 1905—sandwiches and free drinks with the new models

A 1905 4-cylinder Darracq with liberated woman—soon to be a buying force in her own right

The new status symbol on parade—Forrest Park, St Louis, Missouri, c. 1917

This almost mystical concern for the 'people' showed in Ford's determination to democratise the motor-car—for the usual mixed motives, altruism tinged with financial ambition. The price of the Tin Lizzie was more than halved to $375, and now at last a restless people, devoted to speed, had their own transportation to be speedy and restless in.

The car did more to open up and unite America than even the railroad. The Lincoln Highway, the nation's first transcontinental road route, was begun in 1913—partly backed by the car companies. The Dixie Highway, Miami to Michigan, followed. And others. Better roads meant better communication, meant more farmlands, more mines, more new industries. Los Angeles became the first city in America built with roads designed for the automobile. But there were, of course, attendant dangers:

Mr and Mrs Wm. K. Vanderbilt Jnr of Newport, R.I. had a narrow escape from death last evening. They were returning home in their locomobile, and were riding at a good gait down Bellevue Avenue. At the end of the avenue is a sharp turn, and it was here that the accident happened.

Mr Vanderbilt veered quickly around the corner and, as he did so, he ran right into a fashionable turnout, drawn by a span of high-steppers.

It looked as if a collision was inevitable, but Mr Vanderbilt was quick to act, and the machine was brought to a sudden standstill. At the same time, the horses reared up, pawing the air excitedly. The driver saw that he must keep the horses in the air until Mr Vanderbilt could back out from under them. Mr Vanderbilt soon had his machine at a safe distance, but none too soon, as he had hardly got from under the horses when they came to earth.

When Mr Vanderbilt turned to his wife to joke away the fright, he found that she had fainted and sat limp by his side. She was taken into a cottage and was soon revived.[116]

And there were unexpected linguistic problems:

Syracuse Dec. 21, 1900.

Editor, *Horseless Age*.

We are, as far as I know, in need of a word or term to properly and briefly designate the place where we store or keep the automobile. For want of a better name I call my automobile stable or storehouse the Autogorium. In our language, we have auditorium, where people hear things; emporium, where people do trading; natatorium, where people learn to swim, etc. Autotorium would be, perhaps, more in keeping with roots and derivations, but Autogorium would be more euphonious, and would not sound so much like auditorium.

Sincerely,

Gregory Doyle.

The automobile was the essence of the American dynamic and the apotheosis of mass-production. The coming of age. And there were few people who really doubted its benefit to society.

The improvement in city conditions by the general adoption of the motor-car can hardly be over-estimated. Streets clean, dustless, odorless, with light, rubber-tyred vehicles moving swiftly and noiselessly over their smooth expanse would eliminate a greater part of the nervousness, distraction and strain of the modern, metropolitan life. [1899][117]

The city of 1900, of course, was very different from the city of a hundred years earlier. It had grown and kept pace with technology: technology fashioned its shape and its life-style. In two hundred years America had transformed itself from a rural to an urban society. In 1776 some four per cent of the population lived in 'urban' areas. In 1976 the figure is approaching seventy-four per cent.

If we consider that not eighty years ago the place where Philadelphia now stands was a wild and uncultivated desert, inhabited by nothing but ravenous beasts and a savage people, it must now certainly be the object of everyone's wonder and admiration.[119]

That observation by Andrew Burnaby, an English clergyman, was based on his visit to Franklin's Philadelphia in 1759. By 1800, America's biggest city had a population of 73,000 and, by 1850, 450,000. Boston grew from 38,000 to 212,000, New York from 63,000 to 700,000. As immigration increased and the movement west began, a city would spring up in the middle of nowhere for no better reason than that the natural contours of the ground, or perhaps proximity to a river, made the place a convenient overnight stop for wagons.

What a wonderful city Buffalo was in those days! What vast wealth its citizens had accumulated in a few months preeeding! 'Dollars'— 'Dollars' was on the lips of everyone. 'Colonel P. made 50 dollars yesterday'—'Deacon S. cleared 20,000 by that speculation'—'I will give you 30,000 dollars for the lot.' Such sounds rang in my ears at the hotels, in streets and on the docks.[120]

Buffalo didn't exist in 1800. Fifty years later it had 42,000 people.

Chicago, the 'Wonder of the West' or, according to your opinion, 'that brutal network of industrial necessities', grew in sixty years from one hundred people to one million. Los Angeles, in boom state California, in the 1880s grew from ten to eighty thousand. A case of instant city. By 1901, the total population of America was forty million, about the same as Great Britain.

Mid nineteenth-century America saw the growth of the city as a sign of progress.

Broadway, New York, 1859—photograph taken by William England for the London Stereoscopic Company. Horse buses, Havana cigars and no high-rise buildings.

The Brooklyn Bridge under construction in 1881. One of the triumphs of American engineering, the centre span was the world's greatest at 1595 feet.

The proportion between the rural and town population of a country is an important fact in its interior economy and condition. It determines, in a great degree, its capacity for manufactures, the extent of its commerce and the amount of its wealth. The growth of cities commonly marks the progress of intelligence and the arts, measures the sum of social enjoyment, and always implies increased mental activity. [1843][121]

Once the American System of Manufacture became established in the early 1800s, the rise of the city was inevitable. The system demanded centralisation.

Fifty or sixty years ago, small manufacturing establishments in isolated situations and on small streams, were scattered all through the Eastern states. The condition of trade at that time rendered this possible. Now they have almost wholly disappeared, driven out by economic necessity: and their successors are in the cities and large towns [1895][122]

Other contemporary commentators talk of 'industrial convenience' and cities as 'matters of economy', but the shift away from the countryside was not universally welcomed. And never had been. Thomas Jefferson:

I view great cities as pestilential to the morals, the health and liberties of men. . . . When they get piled upon one another, as in Europe, they will become corrupt, as in Europe.[22]

The city seemed a betrayal of much of what early America stood for. It robbed a person of his independence and self-sufficiency, forced him into complete reliance on other people for services like food and water; his identity was merged with others, and overwhelmed. And whilst the country produced a healthy body and a healthy mind, the city did neither.

. . . where pleasure wreathes perennial flowers, and magnificence runs wild with varied forms; here, in sad refutation of utopian speculation, the leper crouches in dumb despair; the beggar crawls in abject misery, the toiler starves, the robber prowls, and the tenant-house—home of all those outcast human beings rises in squalid deformity, to mock civilisation with its foul malaria, its poison-breeding influences, its death-dealing associations.[118]

It could be Dickens on the slums of London; instead it's an official report of 1857 on the slums of New York. The attempt made by men like Lowell to avoid the Dark Ages of European industrialisation had, apparently, failed. But William Chambers, an English visitor in 1853, saw a completely different New York:

As a great emporium of commerce growing in size and importance, New York offers employment in a variety of pursuits to the skilful, steady and industrious and on such terms of remuneration as leaves little room for complaint.

The city brought prosperity and wealth to most American people in a way that had been impossible in the country.

For the greater part of its length, [Broadway is] a series of high and handsome buildings . . . these edifices are more like the palaces of kings than the palaces for the transaction of business.

A. T. Stewart opened one of these 'palaces' in 1846—America's first Department Store. The unmistakable outward sign of the consumer boom produced by mass-production. But only 'like' the palaces of kings, as this was the Republican United States. From its inception, the American department store was a 'People's Palace' and the exclusivity of the European store was carefully avoided. 'Goods suitable for the millionaire at prices in reach of the millions' were set out on large counters for all to see and for all to handle.

The range, even in 1846, was wide, and it got progressively wider: interior-sprung mattresses (1830), sewing machines (1840s), washing machines (1850s), followed by irons and vacuum cleaners. From 1880, ice-boxes disguised as sideboards; by 1900, kerosene refrigerators disguised as wardrobes. Models were continually 'improved' as technology grew. Wood-fired water heaters gas-fired, then electric. Flatirons warmed over a naked flame became steam irons, gas irons, electric irons. Later, as in the automobile industry, models were updated to overcome consumer boredom with a mass-produced product. 'Updating' often meant just a change of colour, but it made the previous year's product obsolete and kept sales buoyant as Americans increasingly demanded possession of 'the latest thing'.

Broadway, New York, c. 1886—horse-drawn trams, telegraph poles and a maze of overhead wiring

Chester Bullock's Fancy Goods Store (left), 501 Broadway, in 1870—lit by gas

Orchard Street on New York's lower East Side in 1898 (below)—some electric street lamps and signs in Hebrew

Montgomery Ward's mail order business was housed above a livery stable on Kinzie Street, Chicago, in 1874

Parcels awaiting dispatch from the shipping room at Montgomery Ward's Chicago House, 1913

As demand increased, so did the number of stores, bringing about a new concept in mass-marketing—the chain store—designed for the lower end of the consumer market, as the 5 and 10 cent store of F. W. Woolworth. Mass-production brought prices tumbling; bulk buying brought them tumbling further. And large display windows—made possible by technical improvements in the manufacture of glass—brought the customers in. From one store in 1880, Woolworth expanded to fifty-nine stores in 1900, and then became one of the first to export American domestic marketing techniques to the rest of the world. The red and gold facia of the American store on Main Street became as familiar in the British High Street.

To keep the consumer treadmill turning, the American banks pioneered credit in 1860. Another American invention, the instalment plan, became as widespread in 1910 as the motor-car it was intended to help pay for. The demand was there, the supply was there, only the ready cash was missing. Various cash-substitute payment plans provided the answer—to the mutual satisfaction of manufacturer and consumer. And to bring the department store and the fruits of mass-production to rural America, the Mail Order catalogue. It came out of Chicago, like so much else that's truly american American created in 1872 by Montgomery Ward:

'The Cheapest Cash House in the country',
'Payment Only if Satisfied'.

There was much opposition from local traders and the local press:

Grangers Beware! Don't patronize Montgomery Ward and Co.—they're dead beats. Another attempt at swindling has come to light. [*Chicago Tribune*, 1873]

But, by 1895, on a pitch of big choice, cash sales and money-back guarantees, Montgomery Ward had become, in its own terms, 'the Most Complete Store on Earth'. It sold everything, from

Soft Rubber Urinal Bags. French pattern. Day and Night use for male. Each 0.95 dollars.

to

A Windsor Organ. Chapel Style. Including instruction book. Each 37.00 dollars. Weight, boxed, about 300 pounds. We sell hundreds upon hundreds of organs and scarcely ever receive a complaint.

and:

Indiana Piano Box buggy for 45 dollars plus trimmings
Angora Lamp mat 0.20 dollars
Wind mill. Price complete, except tower, 30.00 dollars

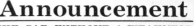

Or any number of drills, ploughs, rakes, carts, forks, pullies
and clothes, as well as ladies' scent. The idea caught on. Richard
Warren Sears' 'Cheapest Supply House on Earth' began rival
business in 1887. It soon offered over 10,000 items, and by the
end of the century had overtaken the sales of Montgomery Ward.
Today—a measure of the firm's success—the Sears Building in
Chicago is the tallest building in the world.

Department stores were pioneers in more than selling. Paper
bags were a newly-introduced convenience in 1867; the cash
register the convenience innovation of 1871. And the buildings
themselves: often the largest in town, they were also often the
most advanced in terms of construction. The department store
was the first rung of the ladder that climbed to the skyscraper.

Joseph Paxton, with his London Crystal Palace of 1851, had
shown the world how to build with iron and glass. In 1862,
James Bogardus, 'builder of cast-iron houses and Manufacturer
of the Eccentric Mill', showed Americans how to do the same.
A. T. Stewart's business had been so successful that he needed a
new store. Bogardus gave him one. Built on a cast-iron frame it
had a large cast-iron fireproof façade and innovations consisting
of an open-floor plan, high ceilings and large plate-glass windows,
to let in daylight as well as display goods. The biggest iron construction in the world, it stretched on New York's Broadway from
9th to 10th streets and was eight storeys high. Bogardus, enthusiastic for the new method of construction, predicted buildings of the future not eight storeys high but *ten miles*.

It wouldn't be possible without steel. The American iron and
steel industry had come a long way since the expulsion of the
British had enabled them to make their own nails, and, since the

opening of their own foundries in 1820, their own guns. Now, in 1860, the new hero was William Kelly. Kelly, like William Bessemer in England, was experimenting with methods to convert iron into steel. Both men discovered the process of strengthening iron by blasting cold air into the molten ore to burn out impurities. Kelly built the first Bessemer Converter in the US in 1862. No commercial use was made of it, and steel was not available for Bogardus. But the demand for steel for the new machinery and for the burgeoning railroads produced the first steel rails to be made in America. With customary rejoicing:

This country has thrown off another shackle that has hitherto bound it to England. Bessemer steel rails may now be purchased from the Cambria Iron Works. [*Pittsburgh Chronicle*, 1867][2]

By 1876, the British steel industry had a dangerous rival.

Dear Sir,
  Whilst your metallurgists, as well as those of France and Germany, have been devoting their time to the discovery of new and fancy processes, we have swallowed the information so generously tendered through the printed reports of your Institute and have selfishly devoted ourselves to beating you in output.[95]

The building material essential for the high-rise was now readily available.

  The skyscraper, an all-American concept, was naturally the product of all-American Chicago. Plus a combination of Act of God in the form of natural disaster, Act of Man in the form of rising land prices, and Act of Faith in man's continued determination to laud it over nature. In 1871, Chicago was still a ramshackle of wooden houses. Some stores and factories were of brick, but very few. On 8 October the city burned to the ground. The McCormick Reaper Works went with it and, for a time, the aspirations of Montgomery Ward. One hundred thousand people were without homes. Nearly two hundred million dollars worth of property was destroyed. But in the process of reconstruction the city doubled in size, then doubled again—and evolved both a new system of building and a new style.

  The Apartment House was a product of pre-fire Chicago. Harriet Beecher Stowe called it, in 1869, a method of 'economizing time, labour and expense by the close packing of conveniences'. The 'all under one roof' manufacturing principle had spread from the factory to the department store and now to the home. With the increased use of steel after the fire, the apartment house began to grow taller. From a successful unit for communal living, it was quickly adapted as a successful unit for communal

The Pennsylvania Steel Company – a view of the Bessemer plants, mills and shops from the Pennsylvania Railroad, c. 1875. The Bessemer Steel Converters at work, 1886 (below)

The Auditorium Hotel by Louis Sullivan, Chicago, 1887 (left)

The Reliance Building by Burnham and Root, Chicago, 1895

office working. The Home Insurance Building (1884) was one of the first: steel frame, walls freed from their supporting function and turned into windows, as in the department stores. In 1885, the Tacoma Building pioneered a new type of foundation. Instead of the brick and concrete overall platform, support for the columns of the steel frame was gained by sinking piles in concrete —a technique learnt from the engineers who built the Brooklyn Bridge, finished in 1883. Along with that innovation went the necessary inventions of the pile driver, mechanical digger and pneumatic drill. As the buildings went higher, so did the cranes.

In 1887, the Chicago Auditorium Building. 1889, the Leitner Building: 'a giant structure . . . healthy to look at, lightsome and airy while substantial . . . a commercial pile in a style undreamed of . . .'. 1890–5, the Reliance Building: 'a glass tower fifteen storeys high'. 1899, the Carson-Pirie Department Store. By now the movement had spread to New York:

We are getting more accustomed to the lofty structures, and so conventional ideas, born of what we are accustomed to look at, are gradually being modified. . . . It is in the aggregation that the immense impressiveness lies. It is not an architectural vision . . . but it does, most tremendously, 'look like business'.[123]

A technological revolution, but also a visual revolution without equal since the master builders of Europe overawed the population with the towers and spires of the medieval cathedral. Skyscrapers: cathedrals of commerce. A new style catalysed by business 'thrust': quick to build, as functional and utilitarian as the coffee urn. As unafraid to display its working parts as the Corliss steam engine. And beautiful. And, despite the reservations of some critics, architectural. America's first internationally-renowned art form—before the movies, the musical and jazz; before its progeny, abstract sculpture and painting.

In stylistic contrast, Britain was still in a fall-out period from High Victorian Gothic. Her commerce was in retreat. She housed it, perhaps significantly, in buildings redolent of past glories—pseudo-Queen Anne or monumental Roman. Of course, those new rich Americans and those American cities who wanted the prestige of European old age were doing much the same thing—Washington, for example. But Washington is not America.

The high-rise could not have functioned without the invention of Elisha Graves Otis. Otis, son of a Vermont farmer, worked in New York in a bed factory. To move the beds from floor to floor he devised an elevator. There had been lifts before—the cage suspended on ropes and drawn up and down by a counterweight that had to go down as far into the ground as the lift rose above it. It was slow, potentially dangerous as the ropes could snap, and unpopular on both counts. People were willing to risk freight in them but not themselves.

Otis, to speed up the movement of the cage, devised an ingenious system of pulleys, but it put even more wear and tear on the ropes. A safety device was essential. To the walls of the cage

The Woolworth Building, New York, 1913 —American high-rise plus European High Gothic on a scale never seen before

Advertising leaflet, 1855

shaft, he fixed ratchets, to the car itself, teeth. The ratchets and teeth were kept apart by the cage rope holding the teeth back. If the rope went slack or actually broke, the teeth were set free and bit into the ratchets. Otis demonstrated his invention at the New York Crystal Palace Exhibition in 1853. The crowds assembled. Otis hauled himself up with all the panache of the travelling showman, brandished a knife—and cut the rope. Instead of crashing to his death, the lift held—and the crowds were convinced.

The first commercial elevators, hand-operated, were installed in department stores and hotels in 1857 and 1858. They travelled about fifty feet a minute. In 1861, Otis patented his steam elevator; in 1876 he won his prize at the Centennial, and in 1889 installed a hydraulic lift in that earliest of cast-iron skyscrapers, the Eiffel Tower. Fifty people at a speed of four hundred feet a minute. In the same year, the always avant-garde department stores like Macy's installed elevators that were electric. A tailor-made form of transportation was ready and waiting for the invention of the skyscraper.

The great improvement of the lift or elevator [has] added probably 10 per cent actually and much more theoretically to the possibilities of population on a given amount of ground.[122]

OTIS AUTOMATIC
ELECTRIC ELEVATOR
FOR USE IN RESIDENCES

is an addition to the comfort of every member of the house-
hold; and at the same time increases the value and salability
of property more than cost of installation. No house of pre-
tension should be without one. We frequently install elevators
in houses already built. It is not as much of an undertaking
as one might think to thus bring an old house up-
to-date. Write for blanks and specifications.

OTIS ELEVATOR COMPANY,
New York Office, 17 Battery Place

Branch Offices throughout the Country

And now, within a very recent period, three new factors have been suddenly developed which promise to exert a powerful influence on the problems of city and country life. These are the trolley, the bicycle and the telephone. It is impossible at present to foresee just what their influence is to be on the question of the distribution of population; but this much is certain, that it adds from five to fifteen miles to the radius of every large town, bringing all this additional area into new relations to business centres.[122]

To a large extent, the size of an American city was still governed by the distance a human being could comfortably walk to work or shops. About three miles. As work and shops became increasingly centralised in the downtown area, homes began to retreat towards the city limits. Cheap and efficient transport became essential.

For public use there were carriages for the rich, horse-drawn omnibuses for the not-so-rich, and a multiplicity of devices in different cities based on the technology of the railroad.

Even as we write, a comfortable passenger car is running smoothly and safely between Warren and Murray Streets, demonstrating, beyond contradiction, that it is only a question of time and money to give us rapid and comfortable transportation from the Battery to Harlem River. [*New York Evening Mail*][76]

The underground railroad opened on 26 February. It consisted of a cylindrical tunnel along which cylindrical railroad cars were blown by an engine, itself powered by a 100 hp steam engine. Plans for an elevated railroad for New York were drawn up in 1872, but most cities settled for street railways with carriages drawn by horses. But not for long.

In the US, electric trams and railways are becoming common for there are at present 23 towns in which this mode of locomotion has been adopted. . . . It is prophesied that in another ten years or thereabouts, horses on tramcars will have been altogether superseded with benefit to man and beast alike, for the poor quadruped of our streets has no harder work to do than the continually stopping and starting of these heavy vehicles. But we cannot boast that humane feelings have had much to do with the change. The fact is, that while horse-flesh costs about five pence a mile, the electric system is about one penny less. It therefore pays well to be humane in this matter. [*Chambers' Journal*, 1887]

In 1890, to do away with the inconvenience of having a rail laid in the middle of the street, Boston introduced the electric trolley with power taken from an overhead wire. The vehicle was little more than a box on wheels, roofed, but open on all four sides;

Trolley car on the Coney Island to Brooklyn railroad, 1897

New York's first electric taxi-cabs appeared in 1898 (above). An invention of Edison's, they didn't last, and by 1907 were ousted by the internal combustion engine.

with four large bicycle wheels and a fifth steered by a tiller, as in the contemporary Duryea automobiles or the electric cars of Edison. Other cities followed:

The trolleys seem to have created a new patronage of their own. Travel has been stimulated rather than diverted . . . [the number of] people who commuted to Manhattan was greater than the population of a city the size of Cincinnati: 100,000 came by bridge and ferry from Brooklyn: 100,000 plus from New Jersey, and more than 118,000 at Grand Central from Westchester and Connecticut. [*Harper's Weekly*, 1896]

Many of this new breed, the commuter, came all or part of the way by bicycle. Introduced in the 1880s, the bike became a national craze. New, manoeuvrable, fast and cheap, it was free from the destination restriction of rails and, most appealing to Americans with their orientation towards self-sufficiency, it was self-propelled and could be privately owned.

In America, cycles are used by every class of American from physician to telegraph boys; and it is reported that there are 500 lady tricyclists in Washington. Medical men in our midst have been recommending this exercise to ladies, with certain reservations, so that this form of exercise has in all likelihood even a greater future before it. [*Chambers' Journal*, 1888]

Ingenious suggestions were made for bicycle improvements.

A new form of boat, which may be described as a water-bicycle, has recently been tried with success in New York harbour. This curious vessel consists of two cigar-shaped tubes, each 12 inches in length and one foot in diameter connected together by an iron framework. Between the tubes is a light water-wheel which is worked by pedals, the navigator being seated upon a bicycle saddle fixed above the wheel.

'Velocipede'—a song to celebrate the advent of a new national craze

Bicycling on Riverside Drive, New York, 1895, before lady cyclists took to the Bloomer Costume (1899, left)

Although, on the day of the experiment, the wind was blowing hard and there was a choppy sea, the novel boat travelled three miles in forty-five minutes. [*Chambers' Journal*, 1887]

But the bike was most important in America for the invention of Mrs Amelia Bloomer, whose garment for ladies offered a physical release from the restrictions of the corset, and important for the attendant social release for women now allowed a 'respectable' leisure activity to accompany their 'respectable' jobs. The bicycle craze soon blew itself out, but it was the bicycle that first 'directed men's minds to the possibilities of independent, long-distanced travel over the ordinary highway'.[110] In other words, primed people for the advent of the internal combustion engine, which appeared on the streets of New York in the guise of a taxi in 1907, and

'Ladies without Bloomers not allowed on the Beach'—Ocean Grove on the San Miguel River, Pinion, Colorado, 17 July 1897

shortly afterwards as a New York omnibus and a private motor-car. The car had all the advantages of the bike, only it went faster, was more comfortable and, most of all, was a novelty.

The American city, like its European counterpart, was largely a product of nineteenth-century industrialism. A monument to mass-production. Symbol of the new power—gas, electricity and oil. By 1900, it had taken on a recognisably modern appearance: railroads, buses and cars (a few) in the streets; electric lighting, indoors and out; in offices and homes, telephones and typewriters as standard equipment. Plumbing and central heating at high level, air conditioning—first gas, then electric—all introduced in the 1890s. And in the kitchen, the fridge and convenience foods. The last vestiges of the American frontiersman fighting his hostile environment were gone. Mass-production and urban living had made life more comfortable for more people than it had ever been before, anywhere in the world. It was as if another aspect of the American dream had come true.

But the city was doomed. The widespread use of public transport accelerated the transformation from urban living to suburban. The automobile meant work in town, live in the country. The city centre began to decline. The department store declined in the face of the suburban shopping plaza; public transport in the face of private. And, as Henry Ford had made car ownership easy, the decline was greater and faster than in Europe where car ownership, till recently, remained a luxury. Europe today is trying hard to repeat the mistakes America made yesterday.

By 1950, the centres of many American cities were either decayed or given over entirely to commerce and, as living units, dead. America was faced with the fight against another hostile environment, but this time man-made, with its traffic jams, pollution, energy crisis and urban violence. But that was all in the future. In 1900, a unique system of manufacture, unrivalled natural resources and ruthless business know-how kept the skyscrapers of commerce rising. And America took over from Britain the role of Shopkeeper to the World.

WHITE MOUNTAIN

REFRIGERATORS
·1903·

MAINE MANUFACTURING COMPANY
NASHUA·N·H·U·S·A·
ESTABLISHED - 1874·

Chapter 6

'Street of Tomorrow'—General Motors Pavilion, New York World's Fair, 1939

The Business of America is Business

Guglielmo Marconi (1874–1937, Italian father, British mother) brought radio to America in 1899. With its huge and growing population, its thirst for novelty in consumer products, America was now the world's biggest market place. Radio made it even bigger—not only by providing a new consumer product, but by providing a means of promoting others as well.

The search for a method of wire-less telegraphy had lasted over fifty years. In an attempt to find another medium for conducting electricity, Morse had tried water—unsuccessfully. Other men, later, tried air. In 1864, the Cambridge physicist James Maxwell propounded the theory of electromagnetic waves. Waves which occur naturally in space, which travel at 186,000 miles a second and are similar to light waves except that they don't illuminate an object but, instead, can change the object's nature. No one believed him; not even the great Lord Kelvin who, in twelve years' time, would so appreciate the achievements of Bell.

Maxwell's theories were proved correct by practical experiments conducted by the German Heinrich Hertz. Hertz used condensers, in the form of Leyden jars—updated versions of the same device used by Franklin—and showed how an electric spark produced by a charge on one jar induced an identical spark on another jar several yards away at the other end of his laboratory: proof that electromagnetic waves—'Hertzian' waves, radio waves—existed, and might be utilised.

It caused a flurry of excitement.

Sir Oliver Lodge, 1851–1940

'The present', said Sir Oliver Lodge, 'is an epoch of astounding activity in physical science. Progress is a thing of months and weeks, almost of days.'[124] In 1894, Lodge, later one of Marconi's consultants, invented the 'coherer'. It consisted simply of a tube loosely packed with iron filings—so loosely that it didn't conduct electricity. Lodge, researching the behaviour of electromagnetic waves in the air, discovered that passing waves, picked up by an aerial, caused the filings to cohere—to stick together and become a conductor. With a telephone receiver added to the 'coherer', it became possible to hear the waves in the form of a click. It was ideal for the transmission and reception of electrical impulses made intelligible by the long and short of Morse code. Marconi was spurred on.

It seemed to me that if the radiation could be increased, developed and controlled, it would be possible to signal across space for a considerable distance. My chief trouble was that the idea was so elementary, so simple in logic, that it seemed difficult to believe no one else had thought of putting it into practice. I argued there must be more mature scientists who had followed the same line of thought and arrived at similar conclusions. In fact, Oliver Lodge had, but he missed the correct answer by a fraction. From the first the idea was so real to me that I did not realize that to others the theory might appear quite fantastic.[125]

It did, until Marconi made his idea public; then it didn't, and he was accused of plagiarism.

By availing myself of previous knowledge and working out theories already formulated I did nothing but follow in the footsteps of Howe, Watt, Edison, Stephenson and many other illustrious inventors. I doubt very much whether there has ever been a case of a useful invention in which all the theory, all the practical applications and all the apparatus were the work of one man.[125]

Marconi perfected Lodge's device and arrived in England in 1896 to a welcome from the *Strand Magazine*. It was just a year since Roentgen had discovered X-rays. 'Today', cried the *Strand*, 'we are confronted with something more wonderful, more important, more revolutionary, the New Telegraphy.' Marconi gave an interview.

I was sending waves through the air and getting signals at distances of a mile or thereabouts when I discovered that the wave which went to my receiver through the air was affecting another receiver which I had set up on the other side of the hill. In other words, the waves were going through or over the hill. . . . I am forced to believe [they] will penetrate anything and everything.

Q: Then you think these waves may possibly be used for electric lighthouses when fog prevents the passage of light?

Marconi's 'improved' version of the 'coherer'. The tube was filled with iron filings.

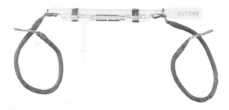

A: I think they will ultimately be so used.
Q: Do you limit the distance over which these waves can be sent?
A: I have no reason to do so.
Q: Then why could you not send a despatch from here to New York?
A: I did not say it could not be done.

The interview concluded: 'He is completely modest, makes no claims whatever as a scientist—but the facts and instruments are so new, that the attention they are at present exciting, is extraordinary'.[124]

One of the most excited men was W. H. Preece, Chief of the British Post Office Electrical Department. What most impressed him was Marconi's innovation of an aerial and an earth. By extending one antenna farther into the air he increased the range of radio waves. By burying the other antenna in the ground, radio waves could be transmitted over several miles instead of several yards. In 1897, Marconi transmitted signals eighteen miles. In 1898, Queen Victoria invited him to put her in wireless communication with the Prince of Wales who was hiding from her on the Royal Yacht. In 1899, signals were transmitted across the Channel. In 1901, they crossed the Atlantic. On 12 December, from Poldhu in Cornwall to St John's Newfoundland, a distance of 2170 miles. The world, and especially America, was agog.

Marconi at Signal Hill, Newfoundland, in 1901 with the instruments used to receive the first wireless signal across the Atlantic from Poldhu, Cornwall

'Caution, Very Dangerous, Stand Clear.' Part of the transmitting station at Poldhu. This station was one hundred times more powerful than any other previously built.

The radio was Europe's first major scientific export to the US in the twentieth century—but it repeated the pattern of the nineteenth. Europe showed how to invent it—America showed how to perfect it and broaden its use. From the start, Marconi conceived radio as a development of the telegraph, extending telegraphy to areas where before it had been impossible—particularly communications from shore to ship. It was on this aspect of radio that Britain, still the greatest maritime power in the world, concentrated. Just one year after his interview with *Strand Magazine*, Marconi was sending signals twelve miles from a British lighthouse to a lightship; the radio link was maintained for two years. In 1899, three warships of the British Navy kept in Morse contact over a distance of almost one hundred miles. Radio stations were erected all along the British coast by the Marconi Company who, by 1902, were handling over six thousand messages a year.

The American Navy put radio on her ships in 1899—for signalling. But the Americans, who tended to be garrulous, and who had already become accustomed to the telephone, were more interested in the transmission of the human voice.

'If anyone hears me, please write to Mr Fessenden at Brant Rock.' Reginald Fessenden's words were picked up by surprised ship's radio officers on Christmas Eve, 1906. Navy wireless stations down the Atlantic coast picked up the message as well. It was followed by a violin recital, a Bible reading which included the words 'Peace to Men of Goodwill' and Christmas greetings to the handful of listeners equipped for continuous wave detection. It constituted the world's first radio broadcast.

Professor Reginald A. Fessenden (1866–1932) of the National Electric Signaling Company was transmitting out of his laboratory at Brant Rock, Massachusetts. He was using a new high-frequency alternator and an umbrella-type aerial on top of a forty-foot tower. Fessenden was the Alexander Graham Bell of radio—the first man to transmit the human voice through the air. He'd already done it in 1900 using an old Hertzian 'spark' transmitter. But the voice could barely be heard. Something more sophisticated and complex was obviously necessary. His answer was the high-frequency alternator, which got rid of much of the extraneous noise that made speech transmitted by 'spark' almost inaudible.

Fessenden, one of the five contenders for the title 'Father of Radio' (Maxwell, Lodge, Marconi and Lee De Forest are the others), made two other major contributions to radio technology. His electrolytic detector converted the waves into a continuous flow, greatly improved reception, and made possible the use of headphones. In 1902, it replaced the 'coherer'. Proceeding from

Reginald A. Fessenden, 1866–1932

there, and with the benefit of his research into the use of high-frequency waves, Fessenden pioneered in 1906 the use of 'heterodyne' reception. This was a system whereby frequencies way beyond the range of human audibility could be used to transmit over greater distances, free from interference. By superimposing one frequency on another, a third could be produced—the one heard by the human ear. It was a system far in advance of its time and, before the advent of Lee De Forest's 'Audion', worked better in theory than in practice. In 1912, the National Electric Signaling company went bust.

Lee De Forest (1873–1961) was almost as prolific with radio patents as Edison was with everything else. By 1906 he'd amassed thirty-four. With De Forest, a Ph.D. graduate from Yale, the 'pure' scientist, long dormant in the history of American invention, began to re-emerge. But like his two major predecessors in the field—almost his only predecessors—Franklin and Henry, he believed the fruits of pure scientific research should be applied.

Signor Marconi,
Washington D.C.

My Dear Sir,

Knowing that you are about to conduct experiments for the US Government in the Wireless Telegraphy, I write you begging to be allowed to work at that under you.

It may be that some assistants, well versed in the theory of hertz waves will be desired in that work. If so may I not be given the chance. . . .

As a young man, you will, I know, fully appreciate the desire I feel, in just starting out, to get a start in the lines of that fascinating field so vast in extent, in which you have done so much.

It has been my greatest ambition since first working with electric waves to make *a life work* of that study.

If you can, Signor, aid me in fulfilling this desire, you will win the lasting indebtedness, as you already have the admiration of,

Your Obedient Servant, Lee De Forest,
190, S. Robey Street,
Chicago. Ill. Sept. 22 1899.[126]

Lee De Forest, 1873–1961

Marconi didn't take up the offer. De Forest, disappointed, wrote years later, 'I was even then convinced that the coherer formed a very weak element in Marconi's wireless telegraph system. . . . It was too slow and complicated.'[128] In 1902, in the best American commercial tradition, and in direct competition with Fessenden, Telefunken of Germany and Marconi, he set up the American De Forest Wireless Telegraph Company. He had the advantage over the Europeans:

The United States Government has recently investigated nearly every system of wireless telegraphy, spending 200,000 dollars in conducting tests and the De Forest system has come out on top.[127]

The government, remembering guns, cannon and steel production, wanted to free themselves from having to rely for wireless telegraphy on European imports. And, in August 1903, experiments conducted with De Forest radio by the US Signal Corps over a distance of 107 miles in Alaska were an outstanding success. It led to what De Forest believed 'was the first wireless system in the world operated regularly as part of a telegraph system handling commercial traffic'.

De Forest was never short of ideas. From the *Chicago Tribune*, 8 January 1905:

WIRELESS MESSAGE SENT
TO A MAN IN AUTOMOBILE
De Forest Company's Experiment
in Michigan Avenue Is Successful
—Will Try Moving Machine Next.

An automobile equipped with a brass pole, from which dangled two wires, drew up in Michigan Avenue yesterday afternoon, and at 4 o'clock received the following wireless telegraph message:

'William H. Ocker, Automobile.—How do you like your first wireless ride? The fire department, steamships, and railways ought to adopt the same method of communication.'

The message had been sent from the De Forest wireless Telegraph office in the Railway Exchange Building. . . . Later in the week experiments will be tried in receiving from an automobile running at speed.[127]

In 1906, De Forest experimented with receiving wireless signals over long distances—like the Atlantic Ocean—by hoisting an antenna on kites developed by aerodynamics pioneer Alexander Graham Bell. But in 1907, De Forest discovered that his co-directors were embezzling from the company. De Forest resigned, and set up the De Forest Radio Telephone Company.

I consider wireless telephony more valuable as a commercial property than wireless telegraphy.[127]

What allowed the switch from the telegraph to the telephone, from Morse code to the human voice, was a patent he'd taken out in 1906 for the 'Audion', a three-electrode vacuum tube, 'perhaps the most important event in the history of electronics'.[127]

The Marconi radio had a two-electrode tube. As radio technology developed, it was discovered that various new substances and new devices could also detect radio waves. Crystals were one.

Fessenden's electrolytic detector was another. In 1903, Marconi's collaborator, Sir Ambrose Fleming, produced the two-electrode valve. Fleming, like Fessenden, had learnt much from practical experience in working with Edison, and his valve was like a small Edison light bulb with a metal filament. When an electric current was passed through the filament—the first electrode—it discharged a volley of electrons at a second electrode, a metal plate fixed at the bulb's apex. Electrons—and not every scientist at this date even accepted their existence—are negatively charged. They would only be attracted to the plate when the plate was positively charged. It meant that the two-electrode valve was the best device to date for converting the high-frequency radio waves to low frequency.

De Forest added a third electrode, '. . . in the form of a grid, a simple piece of wire bent back and forth, located as close to the filament as possible'.[127] He tells his own story in his book entitled, none too modestly, *Father of Radio*. He had a knack for self-advertisement:

I now possessed the first three electrode vacuum tube—the Audion, the granddaddy of all the vast progeny of electronic tubes that have come into existence since.

It was patented on 15 January 1907, Patent No. 841,387, 'Device for Amplifying Feeble Electric Currents'. Both detector *and* amplifier, it used an electric power source to boost the radio waves as they were received. The human voice could be heard loud and clear. And it made wireless communication over long distances, using high frequency as envisaged by Fessenden, not only possible but practicable.

Early in February 1907, I had begun to use my new carbon-arc generator of undamped high-frequency 'wireless waves' as a radiotelephone transmitter for the benefit of any wireless telegraph operators who might hear it . . . so, from both my own lab. and the Telharmonium Building (Broadway and 45th Street) voice and music were being radiated at that early date.[127]

Three months later, the *New York Herald Tribune* reported: 'There is music in the air above the roof of the Hotel Normandy these days.' Radio telephony, like radio telegraphy, was becoming comparatively easy and, in addition, reliable. Much more reliable than American business. De Forest, like Fessenden, went bust. But not before he had scored three impressive firsts: the first installation of the radio telephone in ships of the US Navy in 1908; the first radio programme from the top of the Eiffel Tower, a stunt worthy of Elisha Otis—

De Forest's 'Audion'. 'The grand-daddy of all the vast progeny of electronic tubes that have come into existence since.'

I was officially informed that my radio music had been heard on the Mediterranean Coast near Marseilles, a distance of nearly 500 miles. This was indeed cause of happiness and rejoicing.[127]

And, in 1909, the first broadcast opera. The event was written up almost twenty years later in the *New York Daily News*:

The subject of the first broadcast experiment was Signor Caruso. Mr De Forest obtained permission from the management of the Metropolitan Opera House . . . and on the roof of the building he placed a small radio transmitter and ran the necessary wires to a microphone which was placed on the stage.

The opera was Cavalleria Rusticana. Caruso was close to the microphone and a party of Mr De Forest's friends had been placed in Newark, New Jersey, some thirty miles from New York and various wireless operators were listening on ships in the harbour. The reports of all of them were favourable and although the matter was referred to in the newspapers no-one realised the significance of the event. [25 June 1928]

In March 1897, in his interview for the London *Strand Magazine*, Marconi had been asked in what other ways he expected his invention to be used. His reply was significant: 'The first may be for military purposes in place of the present field telegraph system.' From World War I science, and particularly the technology of communication, received its customary war-time boost. Radio became an indispensable weapon for both sides. Fleets and armies were commanded, controlled and manoeuvred by telephone even more efficiently than they had been by telegraph in the American Civil War. And the telephone, of course, could go places that wires could not. Germany's attempts to starve Britain into submission by using her submarines to sink merchant ships on the Atlantic run were foiled by the ships sailing in convoys in radio contact both with one another and with destroyers and spotter planes who could warn of the submarines' whereabouts.

Radio also became a weapon of international espionage. The constant flow of information provided by radio intercept meant that enemy plans and enemy strategy could be followed far more comprehensively than was ever possible by balloon reconnaissance à la Lowe, or by the tapping of telegraph cables.

Under the pressure of war, advances in radio technology were enormous. Direction-finding by radio was the result of experiments carried out late in 1913, and resulted in the invention of the directional aerial: a square wooden frame with wire coiled round its four sides. If the wired frame was rotated slowly, radio signals were perceived as louder or weaker, according to the position of the wire in relation to the direction from which the waves were

The radio at war—horse with radio pack and aerial, US Army mobile signal corps, World War I

The radio at war—a 1½ kw Marconi pack set in Mesopotamia during World War I

coming. Signals were strong if the wire was right-angles on; weak if the wire was turned away. It meant that any boat lost in a fog could locate its position by listening to the strength of identifiable call signals put out by wireless stations on shore. It also meant that submarines and spies could be located simply by listening to the strength of their radio transmitters.

Also useful against the submarine menace was the echo sounder. A surface ship would send out sound waves from an echo meter to the seabed, and receive them back on a radio recording device. The time that elapsed between dispatch and return determined the depth: four seconds meant 2000 feet deep—sound travelling at 1000 feet per second, two seconds down and two seconds back. If the sound waves were reflecting not from the seabed but from the less-deep bulk of a skulking submarine, a well-placed depth charge would do the rest. Other detection devices used radio tubes under water, which could 'hear' the sound of ships up to a radius of almost twenty miles.

It was a combination of the echo principle and experiments in directional short-wave radio in the 1920s and 1930s that led to radar. Radio waves transmitted at high frequency, beamed from a rotating scanner like light from a lighthouse and travelling at 186,000 miles per second, rebound to the source of transmission when they hit an obstacle—a coastline, enemy ship, even flocks of birds—and are reproduced on the radar screen in the form of a white outline. Ships or aeroplanes can thereby avoid obstacles or locate targets, the use determined by whether it is peacetime or war. Priority of invention for radar is claimed by scientists in Britain; Americans make counter-claims. Either way, radar was Britain and America's secret weapon at the beginning of World War II, and the most spectacular and ingenious use of radio up to that date.

US armoured vehicles considerably developed since their first appearance in the Civil War: observation automobile and tank, 1917.

When you're feeling kinda blue
and you wonder what to do,
Che-e-ew Chicklets and
Chee-ee-ee-er up!
When you've lost your appetite,
Here's the way to set it right,
Che-e-ew Chicklets, and
Chee-ee-ee-er up![128]

American domestic radio went commercial almost from the start.

Maxwell House, you see, is packed by the *Vita Fresh* method. And Vita Fresh is the *only* method that fully expels flavor-destroying air from the can. Which means that every *bit* of flavor . . . *all* of the mild, mellow goodness of this fine old southern blend, reaches you *full* bodied, *fine*-aroma'd, wholesomely fresh *always*. Just *one* taste of Maxwell House coffee . . . and you'll *know* the difference.

All spoken in a rich southern accent on General Food's 'Show Boat' radio programme on 6 October 1932. It was another forty years before Britain, on an extremely limited basis, followed the American commercial lead. America meant business—Britain didn't.

In the *Jersey City Journal* of 20 February 1909, Lee De Forest made a prediction: 'I foresee the time when news and even advertising will be sent out to the public over wireless phones'.[127] During the early years of the war, De Forest put a new station on the air:

And Columbia was interested as a very cheap sponsor because I was to play each day a goodly number of their new records announcing the title and 'Columbia Gramophone Company' with each playing. Thus I became the world's first 'disk jockey' . . . for this early sire of CBS.[127]

In 1916, De Forest sent out election bulletins:

And on that historic occasion I, as chief announcer, proclaimed at 11 o'clock, just before I closed down the station, the election of Charles Evans Hughes.[127]

Which was both premature and wrong. De Forest, the pioneer, was denied a part in the postwar radio boom. He had his licence revoked for moving his station a few blocks without authority.

Station KDKA began experimenting with radio programmes in late 1919. It operated out of Pittsburgh under ex war-time radio technician Dr Frank Conrad. It claimed to be the world's first. De Forest, with reason, claimed it wasn't. But KDKA's success was little short of phenomenal. People were getting the 'listening' habit—aided by their recent but belated discovery of the crystal set (crystals of carborundum as radio wave detectors had been around for over ten years), and the advice pedalled to radio amateurs by the Marconi Company in their magazine *Wireless World*.

With the discovery of an audience, KDKA, still an amateur station, began regular information and entertainment broadcasting on 23 December 1920. Westinghouse, recognising the market, took over. In Canada, Marconi, which had been broadcasting occasionally since April 1920, went regular in May 1921. Station WJZ, Newark, New Jersey, an offshoot of Westinghouse's KDKA, opened on 3 October in the same year. On 14 November 1922, the British Broadcasting Company Ltd opened up a new

era in Europe with news bulletins and weather reports from station 2LO in London.

As the audience in America increased, so did the number of stations to cater for them: five hundred in five years. And, as the public scrambled for receivers, they gave a giant boost to the new electronics industry. It was to lead to television, transistors, satellites and space.

Westinghouse, like Canadian Marconi, immediately began turning out radios to receive their radio programmes, mostly music and chat, later, comedy shows and Broadway plays. The National Broadcasting Company was founded in 1926 as a subsidiary of RCA, which itself had been born out of the Marconi patents in 1919. The Columbia Broadcasting System followed in 1927. The cost of broadcasting was to be met, in theory, by the company's sale of radio sets. But it didn't work. Soon, advertising space was sold in the radio magazines that gave details and times of forthcoming broadcasts. It was only one step away from direct sponsorship and radio commercials. The move was resisted. 'It is inconceivable,' said Herbert Hoover in 1922, 'that we should allow so great a possibility for service to be drowned in advertising chatter.'[31] The British solved the finance problem by nationalising broadcasting and making every owner of a radio set pay a licence fee. This was impractical in America where the best traditions of private enterprise demanded that each station be privately owned. By 1926, commercial pressures began to prove too strong. Mammon, in the form of direct sponsorship, took over. In the season 1926–7, major network radio programmes included Eveready Batteries' 'Eveready Hour'; Champion Plugs' 'Champion Spark Plug Hour'; Coward Shoes' the 'Coward Comfort Hour' (of 'familiar music'). Amazo Cook Oil sponsored a musical travelogue which featured Don Amazio ('Wizard'), and General Mills had 'Betty Crocker' give food talks.

Commercials soon followed, rather long to start with. Libby's Hawaiian Pineapple told of 'Famous Feasts of History: famous events that have begun at the banquet table'. One episode reported on the Duchess of Richmond's ball the night before the Battle of Waterloo, at which the Duchess—surprise—served 'gloriously sweet tangy pineapple'. Sahara Coal, after several songs of sunshine and happiness from the Saharan singers, urged listeners to 'Insist on patronising a dealer who supplies genuine Sahara coal from the quality Circle mines of Southern Illinois—coal as hot as the suns of the desert.' Long and corny they may have been, but commercials like these gave the biggest boost yet to consumer demand.

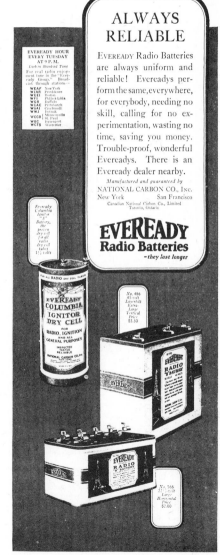

This demand had been steadily increasing throughout the nineteenth century. The American public, grown fat and expectant on a steady diet of new consumer goodies, were eager for more. In the early years of the twentieth century, to cope with their voracious appetite, the major commercial enterprises set up research laboratories, building on the Edison pattern and often employing men with Edison experience. Bell, Westinghouse, Du Pont, Kodak, Standard Oil, General Electric: firms which grew out of and were named after inventors, in most cases long since bought-out or elbowed off the board.

In the past, it had taken an individual inventor, on average, twenty years from the conception of an idea to the point where the idea was developed into a marketable product. For a firm in the manufacturing business in 1910, the delay was impossibly long. The process of invention had to be speeded up. Given the pressures of demand, of mass-production, commercial viability and commercial rivalry, plus the increasing complexity of technology itself, the establishment of the research team was inevitable.

Given proper research facilities and adequate money, the team could cut the invention process—as Edison had already showed—from twenty years to about five. It completely changed the relationship of the inventor to the businessman. The inventor, even the inventor/businessman like Singer or Edison, no longer sponsored industry: industry sponsored and absorbed the inventor. The inventor—though the term itself was now dead along with the old concept of 'invention'—became an employee.

In 1909, Irving Langmuir (1881–1957) joined General Electric. He was one of the first 'pure' scientists employed by a commercial company to do 'pure' research. General Electric's conception of a research lab was somewhat different from Edison's, a fact that Langmuir recognised.

[To] the best type of research man . . . scientific curiosity is usually a greater incentive than the hope of commercially useful results. Fortunately, however, with proper encouragement, this curiosity itself is a guide that may lead to fundamental discoveries and thus may solve the specific problems in still better ways than could have been reached by a direct attack; or may lead to valuable by-products in the form of new lines of activity for the industrial organization.[130]

Edison's approach had been direct. General Electric's, under its director Willis Whitney, was unique in being indirect.

Instead of being assigned to a specific problem, I was encouraged to become familiar with all the work going on in the laboratory and to select a problem which would be of greatest interest to me.[130]

Dr Irving Langmuir and Dr W. R. Whitney at General Electric Co., Schenectady, New York

Langmuir chose to study the problem of getting a better vacuum inside an electric lamp. A better vacuum would, according to 'the universal opinion among the lamp engineers,'[1] result in a better lamp.

Frankly, I was not so much interested in trying to improve the lamps as in finding out the scientific principles underlying these peculiar effects.[130]

There was no direct commercial pressure. Langmuir, for General Electric, was a 'long-term capital investment.'[130]

Several times I talked the matter over with Dr Whitney, saying that I could not tell where this work was going to lead us. He replied that it was not necessary, as far as he was concerned, that it should lead anywhere.[130]

Where, of course, GE hoped it *would* lead, was to a longer-lasting light bulb, a market breakthrough and big profits.

I felt that I didn't really know how to produce a better vacuum and instead, proposed to study the bad effects of gases by putting gases in the lamp. I hoped that in this way, I would become so familiar with these effects of gas that I could extrapolate to zero gas pressure, and thus predict, without really trying, how good the lamp would be if we could produce a perfect vacuum.[130]

After three years' work, Langmuir came up with the answer:

I could conclude with certainty that the life of the lamp would NOT be appreciably improved even if we could produce a PERFECT vacuum.[130]

Hardly the answer GE wanted or expected. Even less expected was Langmuir's further observation 'that it might be possible to obtain a long useful life' for a bulb by operating a filament in *gas* at atmospheric pressure. The suggestion went against all previous teaching. Whitney thought Langmuir was dreaming. But Langmuir's fundamental research into vacuums and the behaviour of gases under different pressures inside lamps, the 'indirect' approach to the problem, was about to pay off—in all senses of the word. On 19 April 1913, Langmuir applied for a patent. It was granted in 1916, and Patent No. 1,180,159, the 'nitrogen filled lamp', not only improved Edison's light bulb and made GE a fortune, it saved Americans millions of dollars in electric light bills.

Langmuir's contribution is so basic that it has survived herculean attempts to obsolete it and the time cannot now be forseen when it may be displaced.[130]

Having studied the problem of vacuums in the electric light bulb, Langmuir went on to use his research and show how an improved vacuum *could* serve radio. The De Forest audion failed to work at high voltages because of the wasting away of the filament and because of the development of what was called 'blue glow'. Langmuir provided the solution: 'It should be possible to obtain large currents in very high vacuum at high voltage'. He invented a vacuum pump, to

create and maintain a vacuum 100 to 1000 times more nearly perfect than any previously used. . . . We also learned how to release streams of electrons into the open spaces of a high vacuum and to control these electrons. Thus the modern science of electronics was born.[1]

Furthermore, he produced a method of eliminating water vapour by baking his vacuum tubes in an oven. Result, the hard valves of the modern radio set. These valves were not outdated until, forty years later, William Shockley and the Bell Laboratories produced one of man's greatest technological achievements, the transistor— that 'little modest crystal that never wears out'. The kind of industrial research begun by GE had been continued by Bell and continues today.

In the early years of GE when the lab was the best in America, it, or more strictly, Langmuir, produced a new atomic theory; research that led to improved X-rays; controlled hydrogen fusion (energy from water); the science of surface chemistry (the manipulation of atoms and molecules leading to a study of enzymes and proteins); and, during World War I, smoke screens and anti-submarine listening devices. Most of these were concepts or inventions that had begun in Europe. In World War II, Langmuir worked on the problems of de-icing planes and, after the war, on cloud seeding.

But arguably the most important 'invention' to come out of GE, in that it had the most far-reaching effects on the economy and on society, was the notion of obsolescence. As expounded by GE's director Willis Whitney:

Large organizations have both an opportunity and a responsibility for their own life insurance. New discovery can provide it.[31]

It meant a continuous flow of new products deliberately designed to outdate what had been produced immediately before. Planned obsolescence, from light bulbs to motor-cars: the new watchword of American industry. A device to perpetuate the system, mass-production and profit—and one day to land the nation in trouble.

The laboratories, what Louis Pasteur called 'these Holy

Dwellings', these 'Temples of the Future', grew big and omnipotent. General Electric grew in ten years from a research staff of ninety-five scientists and their assistants to one of 300. Langmuir, as presiding genius, worked happily inside the new system. De Forest, a loner on the outside, was defeated by it. In 1923, he produced the 'phonofilm', a new method of achieving synchronised sound, sound and picture photographed together on the same print. GE had produced a device too, but De Forest's was better. He tried to sell the system to all the Hollywood studios. There were no takers.

No rush of picture producers to my door. No, Never, Nein und Nimmer. The Publics dunot vant talking pictures.[127]

In 1926, Warners, searching for a sales gimmick, took a risk on introducing 'talkies'. They used Edison's outdated and outmoded method of synchronising the picture with a phonograph. It was a smash hit. Warner's were in the market for a new system —and so was everyone else. De Forest, who should have made a fortune, didn't. The prizes went to the big companies with the big back-ups and big marketing techniques. That they hadn't the best sound system was overlooked, and De Forest was squeezed out.

Jan. 1st 1937. Dr Lee De Forest, radio engineering pioneer has filed a voluntary petition in bankruptcy in the Federal Court in Los Angeles, California, listing 103,943 dollars as his liabilities as against 390 dollars assets. Other assets listed by Dr De Forest include his Hollywood laboratory where he has recently been conducting television experiments, his experimental library and his machinery.

The public notice appeared in the magazine *Telegraph and Telephone Age*. It wasn't quite the end. In 1946, De Forest was awarded the Edison medal 'for pioneering achievements in radio and for the invention of the grid-controlled vacuum tube with its profound technological and social consequences'. De Forest called it his 'most cherished honour—a laurel better late than posthumous'. Thus, the nation's treatment of a man who should have been its hero long before.

The changed role of the inventor/scientist, his changed relationship to big business, is illustrated in the careers of two giants of the new synthetics industry, Leo Baekland (1863–1944) and W. H. Carothers. Baekland, one of the foremost chemists of his age, is, like Howe of the 1840s, the archetype of the 'poor boy makes good', with the additional qualification that he was an immigrant. Born in Ghent, Belgium, on 14 November 1863, he put himself firmly into the American tradition by following Ben

'*Al Jolson is a wireless enthusiast and has three wireless sets in his house.*' Al Jolson, star of '*The Jazz Singer*', the world's first '*talkie*'.

One of the 'profound technological and social consequences' of De Forest's vacuum tube—assembling radio cabinets at the RCA manufacturing plant, Camden, New Jersey, c. 1940

Leo Baekland, 1863–1944

Franklin's dictum that a poor boy should be able to succeed on his own initiative. He paid his own way through university—not a European thing to do, and by age twenty-five was an assistant professor at Ghent University. In 1899 he visited America.

America, where applied science and technology had for so long been the priority, was short of home-produced chemists. To make up for the deficiency, she changed her 1840s' slogan, 'Home for Distressed Workers', to an implied one, 'Haven for Distressed Scientists'.

One of the greatest economic problems of our time is associated with the stream which has been setting westward across the Atlantic and which even now does not seem to have passed its height. The receptivity of the American continent (in respect both of labour and capital) is very great but is not unlimited. Nor is the supply (of both labour and capital) unlimited in the countries of the Eastern hemisphere. We are within sight of consequences which it may be well to consider. [*Chambers' Journal*, 1884]

The Great Brain Drain, which reached a peak in Europe in the 1950s and 1960s, started a good half-century before.

Baekland was persuaded to stay in the States by the offer of a lucrative post as chief chemist in a photographic company. He was thirty-six.

The impact on Baekland of what he called 'Dear America' was enormous. 'My real intense education', he remarked, 'only began after I had left university, as soon as I became confronted with the big problems and responsibilities of practical life'[131] (i.e. the rough-house of American commerce). In 1893, Baekland set himself up as an independent research chemist—and prospered. 'The Land of Opportunity' was living up to its reputation.

I concentrated my attention upon one single thing which would give me the best chance for the quickest possible results.[131]

The result was 'Velox', a photographic paper of 'superior quality' which printed faster than anything else on the market. But, a familiar snag:

I had been too optimistic in believing that the photographers were ready to abandon the old slow process of making photographic prints. I had to find out how difficult it is to teach anything new to people after once they got used to the older methods.

George Eastman had had the same sales problem.

Even my best friends tried to persuade me from continuing my stubborn efforts.

Robert Fulton had been there before. But Baekland solved his problem in the way that Eastman had, by manufacturing for the amateur market—so successfully that Eastman bought him out and Baekland was a millionaire at thirty-seven.

Commit your blunders on a small scale and make your profits on a large scale.

Baekland was becoming more American than the Americans, his aphorisms worthy of Edison himself.

Free of financial worries and the problems of manufacture, Baekland returned to his first love, chemical research. 'Since many years, it is known that formaldehyde may react upon phenol.'[132] Chemists have been beavering away since 1872, but no one had produced anything that was of much use.

And yet I have convinced myself by often repeated experiments that temperatures above 100 degrees centigrade are best suited for the complete and rapid transformation of the substance into a final insoluble, infusible product of exceptionally desirable qualities.

Five years of dogged and persistent experiment had produced what Baekland claimed as the world's first synthetic resin. He named it after himself—Bakelite. In a paper that he read to the New York section of the American Chemical Society on 5 February 1909, he explained his procedure. It reads like a cookery lesson.

'*Old Faithful*', *the original still in which Baeckland conducted his early work on Bakelite resins*

I take about equal amounts of phenol and formaldehyde and I add a small amount of alkeline condensing agent. If necessary, I heat. The mixture separates in two layers. . . . I obtain thus either a thin liquid or a more viscous mass. My starting materials.

Baekland's 'oven' was his own invention,

. . . an apparatus called a Bakelizer. Such an apparatus consists mainly of an interior chamber in which air can be pumped so as to bring pressure to . . . 100 lbs. per square inch [and] temperatures as high as 160 degrees centigrade . . . so that the heated object during the process of Bakelizing may remain steadily under suitable pressure which will avoid porosity or blistering of the mass.

The 'cake' that resulted

. . . is very hard, cannot be scratched with the finger nail and, in this respect is far superior to shellac and even hard rubber as an insulator, and for any purpose where it has to resist heat, friction, dampness, steam or chemicals.

Wood, coated with the substance, produces something:

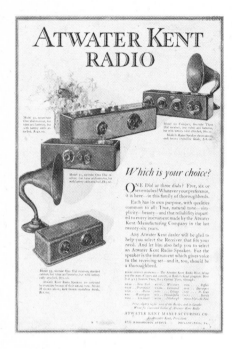

Dr Wallace Carothers with one of his first successes as director of fundamental research at Du Pont. Neoprene—the first commercially successful general-purpose synthetic rubber.

. . . even better than the most expensive Japanese lacquer. . . . But I can do better, and I may soak cheap, porous, soft wood in it . . . transfer the impregnated wood to the Bakelizer and . . . the result is a very hard wood as hard as mahogany . . . proof against dry rot and [which will] bring about some unexpected possibilities in the manufacture of furniture.

Like Bakelite beds. Baekland had a very keen eye for the commercial properties of his new substance:

. . . fastening bristles of shaving brushes . . . pipe stems . . . excellent [for] billiard balls . . . knobs, buttons, knife handles.

In all, some forty-three industries might find it useful. But Baekland was ensured of an even wider application because it had one property not possessed by other 'plastics' like rubber or celluloid.

Several substances which otherwise might be valuable are useless because they CANNOT economically be moulded.

Bakelite could. The new and growing industries—electronics, automobiles—were desperate for a material that combined qualities of resistance with the ability to be moulded into standardised parts that could then be mass-produced. Bakelite was the ideal solution. It replaced rubber as an insulator for electricity and as a material for distributor heads in automobile ignitions. Mixed with sand it made the grinding edge for machine tools. It became the standard material for telephone receivers and the knobs, dials and panels of the new radios. Another industry was born—the synthetics industry—for clothes, televisions, and the nose cones of missiles.

The patents system had often in the past spelt frustration and near-ruin for inventors, Baekland, however, put his faith in the system. 'I have preferred to forego secrecy about my work, relying solely on the strengths of my patents for protection'. For him, it worked well. He won all his legal actions, and Bakelite brought him prestige, happiness and riches in the abundance that the American Dream promised for hard workers with good ideas—a succession of new automobiles, a yacht, a home outside New York, an estate in Florida. It also gave him something of even greater value: 'I was able to work at problems of my own free choice—a free man.' He was one of the last able to do so. It was a freedom that, twenty-five years later, was denied to W. H. Carothers whose researches, unlike Langmuir's at General Electric, were at the behest of the boardroom and whose discoveries became the profitable patents of his company.

Carothers, (1896–1937) was an organic chemist at Yale. In

1927, at the age of thirty-one, he was invited to become head of a new 'pure' science project at the explosives firm of Du Pont. Du Pont had been on the American business scene for over a hundred years and had played its part in fashioning America. In the Civil War it had invented a powder that burned sufficiently slowly to cut down the risk of the giant Union guns bursting in the breach. At the turn of the century, Du Pont powder blasted holes for the foundations of the new skyscrapers and for the New York subway. In 1902, it became one of the first companies to establish a research laboratory in an attempt to diversify its products, capitalise on its knowledge of chemistry, and take advantage of the new consumer markets. For the automobile industry, it produced lacquers, paints, a tar remover (roads were stickier then), and synthetic leather for upholstery.

During World War I, Du Pont supplied some forty per cent of Allied explosives, but with the end of the war and the end of the powder market the need to diversify was even stronger. Remington, now a Du Pont subsidiary, had faced the same situation after the Civil War and bought the typewriter patent. Du Pont now did the same thing and bought up European chemical patents to produce products like cellophane, synthetic rubber and, later, 'Dulux' enamels and insecticides. But in the America of the 1920s, there was waiting to be exploited an entirely new market: women.

Women not only voted now, they bought. And not only for the family, but for themselves. A new purchasing power, new mass fashion stimulated by commercials on radio and glamorous stars on screen. Women in films, women in sport, women in cars, women even in aeroplanes—whatever and wherever, women in work. Wage earners with money to spend . . .

Mary Pickford, the emancipated woman

This company rests on housewives . . . what really makes Du Pont live and breathe is—women.[133]

Feminine interest in gunpowder, of course, was limited. Du Pont, on the General Electric pattern, set up a whole new series of research projects. One of them, fundamental research in organic chemistry, was offered to Carothers. He accepted.

Dr Carothers was among those scientists who had previously shown that it is possible . . . to create long linear molecules in chains.

In 1928, the possibility was still only theoretical.

Molecules are amongst the fundamental 'building blocks' of nature. Invisible under the microscope, they link together to form polymers and, broadly speaking, it is this joining together of molecules that produces growth in plants, animals and man. In

*'Four members of a revue at a San Fran-
cisco Vaudeville House recently evolved the
above method for cutting down the expense
of silk hose. They have adorned their legs
with charcoal, making up their own design.'
Their problem was to be solved finally when
Wallace Carothers invented nylon.*

natural fibres, like silk and wool, these polymers are linear, linked together in a chain formation, very long and very strong.

Carothers, like Langmuir before him, was under no direct commercial pressures. He chose his own field of interest and his own goal—to create giant polymers. This, he believed, could be done, either by simple addition of molecules to form the long chain polymer, or by molecules reacting chemically with each other.

In April 1930, Carothers succeeded in producing, in a molecular still, a strong, viscous substance that contained, instead of the normal short-chain polymers, long chains—as he had anticipated. Super-polymers. One month later, his assistant discovered that from this glutinous mass it was possible to pull out a fibre and stretch it until the molecules in the chain locked firmly into one another and became 'fixed'. The fibre, or filament, was shiny, like silk, and contrary to expectations wasn't brittle at all, but pliable. Carothers had made a fundamental contribution to scientific knowledge, but no obvious contribution to Du Pont's commercial enterprise. The fibre, unfortunately, melted or was easily dissolved. But then commercial enterprise was not, ostensibly, Carother's goal.

In May 1934, Carothers tried again, this time using another process and producing another kind of long-chain polymer. And, instead of stretching the fibre manually, he personally spun it, hot, from the end of a hypodermic needle. The filament cooled into a thin thread. It neither melted nor dissolved. It was what Du Pont were searching for.

Carothers, the 'pure' scientist, retired into the background as the 'applied' scientists took over, producing, in 1935, hundreds of different super-polymers from basic raw materials as cheap and plentiful as coal tar, air and water. Polymer 66 was chosen as the most promising for commercial exploitation and, in 1938, Du Pont presented their discovery to the world. Newspapers carried banner headlines: 'New Silk made on Chemical Base Rivals Natural Product.' But Du Pont had a marketing twist.

A new artificial silk, superior to natural silk or any synthetic rayon in its fineness, strength and elasticity.[1]

The key word was 'superior'. Other artificial products had been marketed as substitutes with the implication that they were inferior to the natural thing. Rayon had suffered in this way. But Nylon was *superior*, and was to sell at a pretty superior price.

. . . lustrous and silky in appearance . . . only one-tenth the diameter of natural silk filament [but of a] tensile strength equal or better than that of silk.[1]

At the New York World's Fair in 1939, Du Pont's 'Wonder World of Chemistry' Pavilion displayed nylon as the bristles for hair brushes, toothbrushes, sanitary brushes; as strings for tennis rackets, line for fishing rods, ropes, upholstery material for furniture, bearings for machinery—soon for parachutes—but immediately for *stockings*. Giant legs, thirty-five feet high, advertised them. Betty Grable wore them. American women fought for them. Long queues (Americans read 'lines') formed round the block when a shop received a consignment. In the first year of production, Du Pont sold sixty-four million pairs of nylons.

As London suffered the bombing and the blackout, Vera Lynn sang 'When the Lights Go On Again'. Broadway sang—'When the Nylons Bloom Again'. For the love-starved girls of Europe, which was the greater attraction—the US serviceman or the nylons in his pocket? Nylons, in many instances, were more valuable currency than the dollar or the pound.

Nylon was an enormous scientific breakthrough, the result of the American dynamic—the same dynamic that had opened up the West—being diverted into an exploration of levels of molecular and subatomic structure. As the geographic frontiers had fallen, so, now, were falling the frontiers of science. But the scientific breakthrough was also a great political coup. Just as Du Pont's gunpowder and chemical research in World War I had relieved the US of dependence on the German chemical industry, so nylon defeated the Japanese silk monopoly. America self-sufficient—the pursuit of that aim continuing over a century after it had begun. But most of all, nylon was a commercial success. Eleven years of 'pure' research and four of 'applied' cost Du Pont a reputed $27 million. It was one of the luckiest, shrewdest, cheapest investments in the history of commercial research: spend a fortune, make a fortune. And a triumphant success for the research laboratory system.

Five hundred patents were awarded to employees of Du Pont, many to Carothers. The most famous, Patent No. 2,130,948: Carothers's in name, Du Pont's in reality as Carothers continued to receive his salary as an employee. In 1936 he had the honour of being elected to the National Academy of Science—the first non-academic and 'industrial' organic chemist to be so. But he didn't live to see the enormous success of his research. In April 1937, he committed suicide. Thirty years earlier, Carothers would have joined Whitney, Franklin, Morse, Edison and Bell as national heroes. Today, his name is commemorated by a plaque on a laboratory wall. The name remembered is Du Pont.

Betty Grable in 'Coney Island'

The New York World's Fair 1939—the hard sell of shiny commercialism

Queues stood for hours to get into the General Motors Pavilion

The commercial giants tried to quicken the pace out of the depression—the Ford pavilion

Consolidated Edison equated electric power with the good life. It supplied electricity to the city of New York and the World's Fair.

Du Pont lined up the lovelies and exposed their legs—in nylons

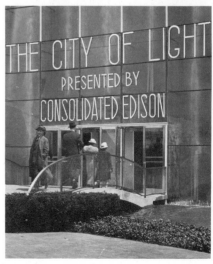

Nylon, like the Pullman car before it, became the symbol of American affluence, American luxury, the 'good life' that the movies portrayed. But there was growing evidence in the 1930s that the price to be paid for mass-production and monopoly was getting higher.

Many facts concur to show that we must look deeper for our salvation than to steam, photography, balloons or astronomy. These tools have some questionable properties. Machinery is aggressive. The weaver becomes the web, the machinist the machine. If you do not use tools— they use you. A man builds a house and now he has a master, and a task for life. He is to furnish, watch, show it and keep it in repair for the rest of his days.

Emerson made these observations in 1870. By 1920, America had roofed the House of Mass Production and Emerson's words were coming true. One cost of keeping the House in repair was the gradual whittling away of what Americans had come to regard as inalienable 'frontier freedoms'. Ford's 'Any customer can have a car painted any colour as long as it's black' might or might not have been a joke, but it pointed up the creeping situation. Cheapness that comes from mass-produced uniformity—but a cheapness at the expense of consumer choice whether of cars, oil brands or telephone systems.

The Trusts had gone, but many of the complaints against them remained. The new large combines provided jobs, but they also restricted job choice. 'Live in Detroit—work at Ford's!' 'When Ford's sneezes Detroit catches a cold.' Not only Detroit, but America as a whole. The car industry as the nation's economic barometer. The system, as Emerson had foreseen, taking over. A new production line, in the name of efficiency and for the sake of profit: the entrepreneur—the worker—the consumer. A momentum that must become self-generating, self-perpetuating, for the House of Mass-Production to remain erect.

On the factory floor, the cost to the worker was denial of his freedom to exercise his American initiative—there could only be one way to do something. And, after the novelty of high wages had worn off, the onset of job dissatisfaction.

I have been told by parlor experts that repetitive labour is soul destroying—but that's not been the result of our investigations. I have not been able to discover that repetitive labour injures a man in any way.

That was the considered opinion of Henry Ford. But the pigeons of the *Principles of Scientific Management* were coming home to roost. And stayed roosted, even for the Ford workers of 1970.

You look around and you ask 'when am I going to be anything other than a clock number?'[135]

Lucy Larcom, 'heedlessly crushed in the strife' at Lowell, had seen it a century before when the system began. Now, in 1932, the worker was crushed even harder between the forces of entrepreneur and consumer. But all three were victims of the system. A scenario:

WOMAN CONSUMER   Dear sir. I felt I just had to write to tell you about your new foundation garment. The colour's right. The fit just perfect.

HENRY FORD   The average worker, I'm sorry to say, wants a job in which he does not have to think.

FORD WORKER (1970)   It slows your thinking right down.

WOMAN CONSUMER   I can honestly say I have never had such uplift at such a reasonable price.

HENRY FORD   They want to be led. They want to have everything done for them and to have no responsibility.

FORD WORKER   There's no change. It wears you out. There's no need to think. It's just a formality.

WOMAN CONSUMER   Please let's have more.

Years of the Depression and labour unrest led to violence at Ford's in 1937. Five thousand demonstrators, and the police opened fire. Four marchers were killed and fifty taken to hospital. 'Communist-inspired big-city demonstrations of the unemployed' was the Ford explanation. But the democratic 'cheap cars for everyone' aim of the early Ford era seemed less democratic in the economic climate of the 1930s. The 'system' was the popular evil and Ford its protagonist.

The simplest way to assure safe production is to keep changing the product—the market for new things is infinitely elastic. One of the fundamental purposes of research is to foster a healthy dissatisfaction.[31]

This from Charles Kettering, of the research laboratory at General Motors, 1932. Despite the Depression, planned obsolescence was now as integral a part of the system as interchangeable parts and the production line.

Evelyn my dear. I'm so glad you dropped in this afternoon. I'm brimful of news. Oh yes. I must confess I've been shopping again. Not window shopping this time but really buying. Listen, yesterday I saw the most perfect dinner things at Wanamakers in the Costume Jewelry Department. They are large and cleverly designed in the very newest styles. I bought one to wear with my black and white ensemble. It's set with marquesities, onyx and silver. Don't you love it? I wore it last night at the informal dinner party the Mannings gave and the girls

The blockade and part of the picket line at the Ford plant, Dearborn, Michigan, 21 May 1937

were crazy about it. Of course, I didn't tell them I only paid $5.95 for it. Oh, my dear, I must tell you about the perfectly enchanting modernistic necklaces. They're in the same department and this week they are only priced at $3.98. Did you ever hear of such a thing? I never did. Alluring colors, genuine stones and there are a few pieces of silver and old gold wrought in the most intriguing chains and pendants you ever saw. Don't you love to shop these days? So many pretty things and so reasonably priced. When I came home adorned with my afternoon purchases, Julie, my French maid, was ecstatic. [*Radio Commercial for Wanamakers*, 1932]

The consumer, said Ford, was more important than either the entrepreneur or the worker as it was the consumer who bought the product. And, within the rules of the system, Ford was right. 'Machine Men with Machine Hearts', cried Chaplin in *The Great Dictator*.

The 'Big Brother' element at Ford's grew as the business grew.

Mention his name to one of [the workers] and if you had the impression that the Ford plant was a paradise for the working man, that impression will disappear in a welter of picturesque epithets.[137]

The system inimical to the freedom of the individual. In 1940 an employee was dismissed for laughing and causing the production line to be delayed for 'maybe half a minute'.

The visitor is struck by the restraint amongst workers: even in moments of idleness, men stand apart from each other. [*New York Times*]

The auto firms, not only Ford's, violently opposed to unionism. In three years between 1933 and 1936, General Motors spent a million on worker espionage.

The spies and stool pigeons report every action, every remark, every expression . . . no-one who works for Ford's is safe from the spies—from the superintendents down to the poor creature who must clean a certain number of toilets an hour.[137]

Irving Langmuir at General Electric, a scientist who put science first and, as he got older, business first along with it ('I like free enterprise'), still saw himself in the tradition of Franklin, Edison, and of science bringing benefits to 'the masses of our people'. But he also recognised the dangers:

All these changes that have resulted from the application of science have led in recent years to perhaps the greatest revolution in the history of mankind. The rapidity of these changes seems to be continually accelerating . . . of course, such a rapid change has not taken place without bringing many new and serious problems. Much of our present economic maladjustment is due to a failure of our social

The mass-production of the factories had its parallel in the mass-production of Hollywood—except Busby Berkeley's routines were prettier

Busby Berkeley's 'Gold Diggers of 1933',
with its song 'Remember My Forgotten
Man', was one of the few attempts by
Hollywood to comment on the Depression,
with its bread and coffee and soup kitchens
for the jobless

organization to keep pace with changes brought about by scientific progress.[130]

In 1924, the British philosopher Bertrand Russell gave a lecture in America on 'The Effect of Science on Social Institutions'. It was reprinted in the *Survey Graphic* with the editor's note: 'if the reader wishes none of his ideas or feelings disturbed, he is counseled not to read the provocative pages which follow'.

Men sometimes speak as though the progress of science must necessarily be a boon to mankind, but that, I fear, is one of the comfortable nineteenth century delusions which our more disillusioned age must discard.
Science has not only brought about the need of large organizations, but also the technical possibility of their existence. Without railways, telegraphs and telephones, control from the centre is very difficult. . . . In America, a railroad president is almost a monarch . . . [and] even where formal democracy increases, the real degree of democratic control tends to diminish. . . .
Science enables the holders of power to realize their purposes more fully than they could otherwise do. If their purposes are good, this is a gain; if they are evil, it is a loss. In the present age it seems that the purposes of the holders of power are in the main evil, in the sense that they involve a diminution in the world at large of the things men are agreed in thinking good.

But, said Calvin Coolidge, 'the Business of America is Business'.
During the Depression, when business wasn't so good, Hollywood offered a palliative in the mass-produced musical with their mass-produced dance routines. Brilliant by Busby Berkley.

Beauty in the repetitive movement of human machinery. Cellophane girls. Packaged products coming off the chorus line as regular as automobiles or sliced bread. The manufactured dream world, the celluloid wallpaper to cover the cracks appearing in the House of Mass-production.

> A toad, the power-mower caught
> Chewed and clipped off a leg, with a hobbling hop has got
> To the garden verge, and sancturied him
> Under the cineraria leaves, in the shade
> Of the ashen heart-shaped leaves, in a dim,
> Low, and final glade.
>
> The rare original heartsblood goes
> Spends on the earthen hide, in the folds and wizenings, flows
> In the gutters of the banked and staring eyes. He lies
> As still as if he would return to stone,
> And soundlessly attending, dies.
> Towards some deep monotone.[138]

America, perhaps, the toad. But the Show Must Go On. 'The Business of America is Business.'

Muskrat trapper's home, Delacroix Island, Louisiana, January 1941—during the New Deal, Franklin Roosevelt shocked the nation by claiming that one third of its people were 'ill-housed, ill-clad, and ill-nourished'

The good life for the poor whites—home perm adverts, cardboard shoe boxes and Hollywood publicity pictures—to keep out the wet

Graphite Research Reactor at the Brookhaven National Laboratory, Upton, Long Island, New York. Tons of pure uranium metal are introduced into the reactor through holes. A concrete wall, five feet thick, protects the technician from radiation

Kitty Hawk to the Moon

President John F. Kennedy:

I believe that this nation should commit itself to achieving the goal, before this decade is out, of landing a man on the moon and returning him safely to earth. No single space project in this period will be more exciting, or more impressive, or more important for the long-range exploration of space; and none will be so difficult or expensive to accomplish.

The moon was the obvious place for America to go. She had conquered her own continent; she was the world's greatest military and economic power; the only major challenges left were underwater exploration or flight in space. Flight: the greater challenge, the greater release, spiritual as well as physical. Aristotle, the Roman philosopher; Bacon, the medieval English monk; Leonardo, man of the Renaissance—all put forward their suggestions as to how it might be done. But it was not until the eighteenth and nineteenth centuries that men actually got off the ground. Europeans in balloons. They got up all right, but had no control over where they were going or how fast. It was Otto Lilienthal, a German, who first gave manned flight a degree of control.

In 1891, he launched his first glider from the side of a hill. He controlled it by suspending himself from the centre of the plane, legs dangling, shifting his body weight to give balance and direction. As he built up his knowledge of aerodynamics and modified his plane designs accordingly, Lilienthal was able to achieve flights of up to 750 feet. But, in 1896, before he was able to experiment with an engine-powered glider, he crashed. 'Sacrifices', he said as he died, 'must be made.' He wasn't the first, or the only one. In 1899, Percy Pilcher, the British inventor, had his glider towed into the air by a horse team. It crashed before take-off. Pilcher was killed.

There were many false dawns. Hiram Maxim, early motor-car pioneer, gun manufacturer and famous for the Maxim 'silencer', was an expatriate American living in England. In 1894 he undertook to build 'a flying machine that would lift itself off the ground'. Having achieved exactly that—with a 360 horsepower steam engine and a collective wing span of 4000 square feet—the machine collapsed. Maxim abandoned his project.

In America, the most serious pioneer was Samuel Langley (1834–1906), secretary of the Smithsonian Institution.

Otto Lilienthal in flight, 1896, the year he crashed and died

I seldom ever saw a bird flying that I did not think of it [aviation] and even lately I have watched them for hours, trying to understand how they could move about through the air, rising and falling, soaring up and sailing down without any motion of wings.[83]

Langley carried out his research with models made of wood and with propellers driven by rubber bands. To give the models more thrust he installed a steam-engine. Enthusiastically watching was Professor Alexander Graham Bell.

Langley's flying machines—they flew for me today. I shall have to make experiments on my own account. Can't keep out of it. It will be UP with us some day. . . . I have not the shadow of a doubt that the problem of aerial navigation will be solved within ten years.[83]

Bell made his own rather erratic forays into aviation, experimenting with gunpowder rockets for propulsion.

The more I experiment, the more convinced I become that flying machines are practical.[83]

In 1896, he was invited to Langley's specially constructed houseboat on the Potomac River to watch Langley's latest sixteen-foot pilotless model catapulted into the air. Steam-propelled, it flew for almost a mile.

Hiram Maxim's flying machine built in 1894—a contemporary model of the 7700 lb machine

. . . an enormous bird soaring in a great spiral to a height of a hundred feet. . . . No-one could have witnessed these experiments without being convinced that the practicability of mechanical flight has been demonstrated.[83]

Langley, after these model experiments, was ready to retire. He was sixty-two. But US President McKinley, envisaging the machine's war-making potential, budgeted $50,000 and requested

Langley to construct a plane that could be piloted. In 1901, he
built a quarter-size model powered by a 52 hp internal combustion
engine—the world's first petrol-powered aeroplane, though
Langley, confusingly, insisted on terming it an aerodrome. On
7 October 1903, the first trial flight of the first full-scale piloted
plane. The aerodrome to be catapulted into the air by a spring
from the launch pad on top of Langley's Potomac houseboat. An
extraordinary example of Yankee know-how or, at least,
'initiative'. Result—newspaper headlines:

*Samuel Pierpoint Langley (1834–1906)
and his 'Aerodrome' prior to launch off
from the houseboat roof, Potomac River,
7 October 1903*

> Professor Langley's Airship Goes to River Bottom.
> Langley's Aerodrome Swoops into Potomac.[2]

Manly, Langley's assistant and pilot, managed to survive.

On 8 December, Langley tried again. And failed again. The
launch pad fouled the aerodrome during take-off, and Langley's
hopes, along with his machine, lay broken at the bottom of the
river. Langley, so close to success, gave up. Nine days later, the
Wright Brothers made the world's first controlled, powered flight
at Kitty Hawk Beach, North Carolina.

At Langley's funeral in 1906, Bell argued that 'his flying
machine never had an opportunity of being fairly tried'[83] but
Langley's practical experience of control of a full-scale machine,

and of the behaviour of a variety of structural designs, was limited, He did not do gliding experiments like Lilienthal, and his technique of catapulting a launch was, in whatever terms, risky.

Langley and Lilienthal illustrate two distinct conceptual approaches to the problems of aviation. The first—Langley's—held that success could be achieved by first perfecting a stable, pilotless flying machine model that could later be increased in size with the addition of a pilot to work the controls. The other maintained that the pilot should be an integral part of a full-scale machine from the outset, and that only his skill as an aviator would enable the 'unstable' machine to fly. It's the difference between imagining a bicycle solid enough to remain upright by itself and a bicycle that, to remain upright, has to be ridden. In the event, Langley's approach was wrong. Lilienthal's was correct—but he died before it could be proved and the European initiative in aviation was lost.

That initiative was next taken up in a bicycle shop in Dayton, Ohio.

My own active interest in aeronautical problems dates to the death of Lilienthal.[139]

Wilbur Wright (1867–1912) paid tribute. He and Orville (1871–1948) had gone into the bicycle business in 1892, hoping to capitalise on the boom that followed the introduction of the 'safety' bicycle. They were successful. In 1895, they began making their own bikes. In the same year, their embryonic interest in flying was stimulated by the report of Lilienthal's glider experiments. Studying aviation became a hobby. Within four years it grew into a serious study. In 1899, Wilbur Wright wrote to the Smithsonian:

I have been interested in the problem of mechanical and human flight ever since, as a boy, I constructed a number of bats of various sizes. . . . My observations since have only convinced me more firmly that human flight is only a question of knowledge and skill, just as in all acrobatic feats.[1]

He asked for technical papers and, in return, received works by the Frenchman Chanute, then resident in the US, and papers by Langley, Lilienthal and Leonardo. But the Wrights progressed not by the collective observations of other men but by their own.

My observations of the flight of buzzards, lead me to believe that they regain their lateral balance when partly overturned by a gust of wind, by a torsion at the tips of their wings.[139]

This important observation was reflected in the Wright's first mini biplane, built and flown on the edge of town in August

Octave Chanute preparing for a flight in 1896. Chanute helped and encouraged the Wright Brothers

1899. The wings of this 'kite', with a span of only five feet, were twisted or 'warped' for 'lateral stability' by controlling cords which could also shift the wings, individually, forwards or backwards. It meant that if one wing of the plane began to dip, the machine could be righted by twisting the wing to an angle on to the wind to give it more lift. From these experiments, the Wrights made a crucial discovery: that what enables a plane to fly is not the engine or the propeller, but the shape and design of the wings.

During the summer, the bicycle builders began the long process of turning themselves into aviation engineers. In September, Wilbur wrote to his father:

I have my machine nearly finished. It is not to have a motor and it is not expected to fly in any true sense of the word. My idea is merely to experiment and practice with a view to solving the problem of equilibrium.[1]

This, the Wrights' first full-scale biplane, had a wing span of seventeen feet.

We were resolved to try a fundamentally different principle. We would arrange the machine so that it would not tend to right itself.[2]

They made it deliberately *un*stable, unbalanced, in order to learn how to control it, to learn the principles of aerodynamics in the way that Lilienthal had begun. Like the earlier model, this machine incorporated the wing-warping technique, had a front 'elevator' for directing the machine up or down, and provided for a pilot to lie prone, head down, in the centre of the machine. It could be flown either as a box kite controlled by cords from the ground, or as a glider.

Orville (left) and Wilbur Wright

The first trials were carried out at Kitty Hawk, North Carolina, a deserted beach recommended by the Washington weather bureau as the windiest place on the east coast. The wing-warping proved an enormous technological advance on Lilienthal's method of retaining control by the swinging movements of the pilot's dangling body. The 'elevator' was a success too. But the summer's total 'flying' time was only twelve minutes.

Bi-plane mark 2 was flown at Kill Devil Hills, four miles south of Kitty Hawk, in the summer of 1901. Wing-span was increased from 17 to 22 feet; wing area from 165 to 290 square feet. The wire cords to control wing warping operated from a hip 'cradle' in the plane's centre. As the pilot shifted his weight to balance the machine, the cords fixed to the cradle automatically warped the wings. The first glides achieved distances of up to 315 feet in nineteen seconds. Hopes were high. But it was soon obvious that the camber on the wings was unsatisfactory.

Wright glider No. 2 being flown as a kite at Kill Devil Hills, North Carolina, 1901. Wilbur has his back to the camera.

The lift is not much over one-third that indicated by Lilienthal [and] we find that our hopes of attaining actual practice in the air are decreased by about one-fiftieth of what we had hoped.[1]

The wing-warping didn't work as it should. The wing showing the greater angle to the wind, instead of rising, sometimes fell. Incredible disappointment and back to the Dayton drawing-board.

Not within a thousand years will man ever fly.[140]

Wilbur Wright was for giving up. But Chanute, now nearly seventy and himself no small contributor to aviation technology, did his greatest service to flying by persuading the brothers to go on.

Having set out with absolute faith in the existing scientific data, we were driven to doubt one thing after another, until finally, after years of experiment, we cast it all aside and decided to rely entirely upon our own investigations.[16]

During the winter, the Wrights, still running their bicycle business, conducted extensive tests. Amongst them, tests to determine the behaviour of two hundred types of wing surface in a small wind tunnel, by measuring the wind pressure caused by curved and plane surfaces, angles, areas. Orville's results were extraordinarily accurate and gave the brothers the data that would, eventually, enable them to fly.

Biplane mark 3, incorporating the results of the brothers' winter investigations, had a wing-span boosted to 32 feet, a wing area of 305 square feet, a wing camber of 1 in 24 to 1 in 30—as opposed to the previously unsatisfactory 1 in 12—and a new device, a tail. This was in the form of a tall, rigid double-finned, lever-operated rudder to counterbalance the effects of warp drag on a dipped wing, which prevented the wing from lifting as it

Wilbur piloting glider No. 3 in 1902. The double-finned fixed rudder has been replaced by a single fin controlled by chords.

should when the plane was banked. It all worked—except the rudder, which occasionally put the plane into a spin. This, the last major control problem, was overcome by replacing the tail with another that was hinged, single-finned and linked by cords with the warp control. It was this successful co-ordination of wing-warping for lateral stability, a front elevator for climbing or diving, and a rudder for turning, that made the Wrights' third biplane the first fully practical glider in the history of flight.

During the five-week test period at Kitty Hawk, the brothers made over a thousand glides, many over distances of more than 500 feet, the longest $622\frac{1}{2}$ feet, in winds of 35 mph. That they could achieve control and stability in such high winds is a mark of their achievement.

Patent No. 821,393. FLYING-MACHINE.
Filed March 23rd. 1903.
To all whom it may concern:
Be it known that we ORVILLE WRIGHT and WILBUR WRIGHT, citizens of the United States, residing in the city of Dayton, county of Montgomery, and State of Ohio, have invented certain new and useful Improvements in Flying Machines. . . . Our invention relates to that class of flying machines in which the weight is sustained by the reactions resulting when one or more aeroplanes are moved through the air edgewise at a small angle of incidence, either by the application of mechanical power or by the utilization of the force of gravity.[1]

One of the most significant patent applications in history—as significant as the reaper, the sewing machine, the telephone and the electric light bulb—and ten years almost to the day as Bell had prophesied.

In the summer of 1903, the Wrights began to build their first powered machine. But the motor problem was not as quickly solved as Wilbur had anticipated. The new, petrol-driven automobile engines were too heavy. The Wrights had to design their own engines, with twelve horsepower and a weight of 200 pounds. They also had to design their own propeller.

On 23 September the machine left Dayton for North Carolina. A biplane, its wing-span was 40 feet 4 inches, the biggest yet, wing area 510 square feet, wing camber 1 in 20. A double elevator in front and a double rudder behind. Two propellers were driven by chain transmission from the engine and the whole machine was supported on an undercarriage consisting of two 'skids' like sledge runners or skis. The plane had its first run from a launch pad consisting of a sixty-foot rail on 14 December. The machine lifted, stalled, and ground into the sand. Wilbur had lifted the elevator too fast. It took two days to repair the damage.

While we had it out . . . making the final adjustments, a stranger came along. After looking at the machine a few seconds, he inquired what it was.[142]

The familiar today was the extraordinary of yesterday.

When we told him it was a flying-machine, he asked whether we intended to fly it. We said we did, as soon as we had a suitable wind. He looked at it several minutes longer and then, wishing to be courteous, remarked that it looked as if it would fly if it had a 'suitable wind'. We were much amused, for, no doubt, he had in mind the recent 75 mile gale.[142]

On 17 December, the brothers tried again. Orville's turn. The wind was strong—27 mph. The cold, biting.

I look with amazement upon our audacity in attempting flights with a new and untried machine under such circumstances. Yet faith in our calculations and the design of the first machine, based upon our tables of air pressures, obtained by months of careful laboratory work, and confidence in our system of control developed by three years of actual experience in balancing gliders in the air, had convinced us that the machine was capable of lifting and maintaining itself in the air, and that, with a little practice, it could be flown. . . .
After running the motor a few minutes to heat it up, I released the wire that held the machine to the track, and the machine started forward into the wind. Wilbur ran at the side of the machine, holding the wing to balance it on the track. Unlike the start on the 14th., made in a calm, the machine, facing a 27 mile wind, started slowly. Wilbur was able to stay with it till it lifted from the track after a forty-foot run. . . . The course of the flight up and down was exceedingly erratic, partly due to the irregularity of the air and partly to lack of experience in handling this machine. The control of the front rudder was difficult . . . the machine would rise suddenly to about ten feet, and then as suddenly dart for the ground. A sudden dart when a little over a hundred

feet from the end of the track, or a little over 120 feet from the point at which it rose in the air, ended the flight.

This flight lasted only twelve seconds, but it was nevertheless the first in the history of the world in which a machine carrying a man had raised itself by its own power into the air in full flight, had sailed forward without reduction of speed, and had finally landed at a point as high as that from which it started.[142]

It was just four and a half years since Wilbur Wright had written to the Smithsonian.

Only three newspapers carried an account of the Wrights' success. Bell didn't believe it till Chanute reassured him. By 1905, the Wrights had built a plane that could take off, bank, land without damage. By 1908, they had produced the world's first practical passenger plane. In their own words: 'The age of the flying machine had come at last'.

The world's first powered controlled flight, 17 December 1903. Orville is at the controls; Wilbur on right. The launch rail is clearly visible.

Mount Wilson Observatory near Los Angeles, California (right). The 60-foot tower telescope dwarfs the long horizontal telescope bought from the Yerkes Observatory in 1905.

The adventure is not theirs alone, but everyone's; the history they are making is not only scientific history but human history. That moment when man first sets foot on a body other than Earth will stand through the centuries as one supreme in human experience, and profound in its meaning for generations to come. [President Nixon, Apollo Mission, 1969]

It was to be sixty years between the launching of the Flyer at Kill Devil Hills and the landing of Apollo on the Moon. During that time it was America's new determination and subsequent achievement in the realms of pure science that led to the technology that made the space programme possible.

America in the latter part of the nineteenth century was still academically and intellectually dependent on Europe. The large degree of self-sufficiency that she had achieved in industry and technology was missing. For research purposes, American scientists still needed to spend time abroad in the intellectual centres of Europe. But after the Civil War American universities began to emphasise scientific research and build up their own departments. The money for doing so was largely supplied by industry: Chicago University was re-endowed in 1892 with money made available by John D. Rockefeller. The final words of Wesley's dictum 'Gain all you can—Save all you can' became 'Give all you can'. And Rockefeller did. American business financed American brains several years before she began to finance her own industrial research projects.

Two men, both associated with Chicago, were to put the US onto the world's scientific research map. George Hale and Albert Michelson eventually reversed the direction of scientific travel across the Atlantic. Instead of American scientists going to Europe, European scientists came to America.

George Ellery Hale (1868–1938) was probably the world's only young astronomer to be given an observatory for his graduation present. He was twenty-two, the son of a man who had made big money out of Chicago's elevator business. Hale's working life was concentrated on studying the sun: photographing it, perfecting a device for studying its chemical composition, analysing the sun spots, and proving that the sun was magnetic. 'The first association of magnetic fields with an extra-terrestrial body.'[145] In 1892, he joined the University of Chicago and set up the Yerkes Observatory, plying his rich friends for funds. In 1903, at Pasadena in California, he conceived the notion of a giant telescope on top of Mount Wilson; in 1908, a 60-inch reflecting telescope; in 1917, a 100-inch. American academic brilliance and American money combined to make Mt Wilson the most famous observa-

George Ellery Hale, 1868–1938

tory in the world and to place the US in the forefront of astronomy and astrophysics. The hard work of the nineteenth century was beginning to pay off as twentieth-century affluence enabled the US to undertake projects that few other countries could afford. A trend that has increased as American affluence has increased.

The working life of A. A. Michelson (1852–1931) was concentrated on accurately measuring the speed of light. He was only thirty when (assisted by a financial handout from A. G. Bell) he invented the Interferometer. This was a device designed to measure distances in terms of light wavelengths. A single beam of light is split into two bands by a mirror; each band travels at right angles an equal distance and then is reflected back and reunited with the other. If the paths travelled by the two separate halves are in any way different from each other, if they have travelled the same distance at different speeds or different distances at the same speed, it will be apparent to an observer, as the two halves of the regrouped beam will be out of phase and interfere with each other, producing bands of light and dark.

Michelson deliberately sent the light bands in different directions in order to make sure that, if the earth was travelling through the ether, as scientists believed, one band would travel *with* the ether current and one against it and so arrive back at the observer's viewing point slightly later than the other. Unfortunately, Michelson could detect no bands of light and dark at all, and so couldn't measure the speed of light. But at least he thought he showed that the ether was not stationary but moved *with* the Earth. 'A result', said Bell, 'in direct variance with the generally received

A. A. Michelson, 1852–1931

theory of aberration,' and which should 'prove to be of much importance.'[83] Which was to prove an understatement.

Michelson's later experiments showed that the ether had no effect on the speed of light at all and, later, eliminated the existence of the ether altogether. 'It was', as Isaac Asimov says, 'undoubtedly the most famous experiment-that-failed in the history of science.' It was Michelson's experiments that provided the information that was to back up Einstein's theory of relativity (that time and place are all relative to the observer). In 1907, he became the first American to receive the Nobel Prize for physics.

By 1927, Michelson, using an eight-sided revolving mirror, had succeeded in measuring the speed of light. In 1933, after his death, the final calculation, based on his own figures, was announced: 186,271 miles per second. (Today the accepted figure is 186,282.) But, equally important in its implications was other work that Michelson did with the interferometer. In 1920, using a 20-inch-long instrument, along with a 100-inch telescope at Hale's Mount Wilson observatory, he measured the diameter of the giant star Betelgeuse: 250 times that of the Earth's sun. That achievement meant that by applying these methods to industry, equipment could be measured with similar accuracy. Screws and bearings could be adjusted to one ten-thousandth of an inch. Industrial precision, absent from mass-production, now became possible: a precision much greater than anything that could have been achieved by the craftsman tradition that mass-production had denied. And a precision without which the technology of space research would be impossible.

That's one small step for a man; one giant leap for mankind. [Neil Armstrong, Commander, Apollo II, 20 July 1969]

In those exhilarating few moments, the whole world was united to watch man realise a dream he'd had from the time he'd first seen the moon.

In past ages, exploration was a lonely enterprise. But today, the miracles of space travel are matched by miracles of space communication: even across the vast lunar distances, television brings the moment of discovery into our homes and makes us all participants. [President Nixon]

American science and technology, its dedicated research effort, and the American way of doing things, were on public display for the whole world to wonder at. It caught the public imagination. But only for the moment. It wasn't enough to dispel the creeping disillusion with science and technology that has been growing for fifty years—a feeling that America, and the rest of industrialised

Apollo II—Buzz Aldrin returning to 'Eagle', the lunar module, after a lunar walk. The American flag 'flies' next to 'Eagle'.

society, had somehow got its priorities wrong. That the rush to the moon was another example of our having forgotten the world.

As Apollo 11 splashed down in the Pacfic on 24 July, only thirty seconds later than predicted, the Poor People's March camped outside the walls of Mission Control. Science, as symbolised by the space programme, was under attack. Within two and a half years, the Apollo programme came to an end. In December 1972, Eugene Cernan, Commander of Apollo 17, spoke to the world from Taurus Littrow on the moon:

As I take these last steps from the surface . . . back home for some time to come . . . I'd like to list what I believe history will record. That America's challenge of Today has forged man's Destiny of Tomorrow.

It probably has. But not necessarily in the way that Cernan predicted. Public confidence and unquestioning faith in a bright technological future has gone; the hopes and aspirations of a century ago turned sour. Why?

The promise of 1876, that technology would benefit mankind, has largely been fulfilled. The American System of Manufacture has produced, in terms of material prosperity, the world's highest standard of living. American medical achievements, from the first anaesthetics in 1846 to Salk vaccine, heart transplants and nuclear pacemakers, have enabled millions throughout the world to live longer and live healthier than ever before. The achievement is not in doubt. But in the light of the exhaustion of natural resources, and population explosion—the physical and social implications— what is in doubt is whether, in the long term, it is the right achievement. The cotton gin parallel: the gin that seemed to save the South, in the long term destroyed it.

Man has used the increased productivity which we owe to science for three chief purposes in succession: first to increase population, then, to raise the standard of comfort; and, finally, to devote more energy to war. Modern industrialism is a struggle between nations for two things —markets and raw materials. . . . Coal and iron and oil, especially, are the bases of power and thence wealth. The nation which possesses them, and has the industrial skill required to utilise them in war, can acquire markets by armed force, and levy tributes upon less fortunate nations.[146]

Russell, in 1924, expressed an extreme minority viewpoint. Fifty years later it is reflected in a general unease of the majority. Rarely specific, rarely well-formulated, but lurking in such phrases as 'too many people', 'too much stress', 'the rat-race', 'the opt-out', 'the drop-out'. The three Ds of modern disillusion with science: it's dangerous, it's dirty, and it's dehumanising.

A more specific sense of technology having become too big to handle—a force out of control. A suspicion compounded by the fact that technology is much too complex for any but the specialist to understand. And the compensation for our unease: that technology was producing the new and wonderful goodies that we were eager to have and were willing to pay for even in terms of restricted choice, is no longer quite the same reality. There was no immediate domestic relevance for space probes. Maybe such a relevance, such a justification, should not be necessary; exploration, research, justify themselves. But the American public, brought up on a diet of innovation, was, perhaps subconsciously, disappointed that a $24 million research programme produced, as far as they could see, just the non-stick frying pan and a variety of techniques for monitoring pulse rates and blood flow. The National Aeronautics and Space Administration—NASA—admit that future rocket probes to Mercury are not economical, not utilitarian, but defend them on the grounds that 'they are cultural; they expand people's minds'.

Mind expansion—or the space race? 'For the peace and hope of Mankind' or a political exercise in ideological one-upmanship, and a blind for military research? In 1908, Prof. Bell looked to the future:

The Nation that secures control of the air will ultimately rule the world . . . [aerial navigation] means an entire revolution in the world's method of transportation—and making war.

Scepticism of the real aims of technology was born of recent history.

Mr Maxim, who is now engaged in constructing a flying machine on a very large scale, upon which he has already expended about £10,000, has remarked: 'If I can rise from the coast of France, sail through the air across the Channel, and drop half a ton of nitro-glycerine upon an English city, I can revolutionise the world. I believe I can do it if I live long enough; if I die, some one will come after me who will be successful if I fail. [*Chambers' Journal*, 1892]

Like Hitler? Or Bell?

I am not ambitious to be known as the inventor of a weapon of destruction but I must say that the problem—simply as a problem—fascinates me and I found my thoughts taking more and more a practical form.[83]

The fascination for the creator—to destroy. Marconi. Lee De Forest:

When the clouds of World War II first lowered in Europe, I began to study night bombing. Quickly, I conceived and sought to patent a self-directing night bomb.[127]

The Wright Brothers claimed their invention as an invention for peace. Aerial surveillance of armies, the possibility of the destruction of cities, meant there could no longer be surprise and advantage in fighting wars. Men would be forced to come to terms—for ever. But although their father was a bishop of the United Bretheren church, Mammon, and double-think won out.

We believe that the principal use of a flyer at present is for military purposes; that the demand in commerce will not be great for some time. It is therefore our intention to furnish machines for military use first . . . a machine capable of carrying two men and fuel for a fifty-mile trip. We are only waiting to complete arrangements with some government.[140]

As they proclaimed peace, the Wrights touted their plans round Europe.

But since the Wrights, no scientist has been his own man. He is the politician's man—one of a team working directly on government research or, like Professor Arthur Galston, a scientist whose work is misappropriated for political-military reasons. In the 1940s, Galston discovered a chemical which activated a plant's ability to make bean pods.

But one never knows what's going to happen to a scientific discovery that one happens to make. When my discovery had been used for constructive purposes, I felt very happy—it produced more soya beans to feed pigs and therefore people . . . [but] I had written in a very obscure paragraph of my Ph.D. thesis that if one used somewhat higher concentrations of chemical spray on the plant, then all sorts of abnormalities would develop—malformation: loss of leaves.

Galston's life-giving chemical of the 1940s became the defoliant of the Vietnam of the mid-'60s. Galston wrote a concerned letter to President Johnson. To no effect. Today, scientists who share the public's disillusion have recently resigned from NASA. The dawning of a new attitude. Marie Curie, who discovered radium in France in 1896, said politics were not her business. Her daughter Irene—whose research work on radium with her husband Frederick Joliot-Curie led directly to the bomb—claimed that it was. The young, especially, are rejecting science. 'All they see,' said Galston, 'is its end products destroying men and destroying environments.' Bob Dylan's song 'With God on Our

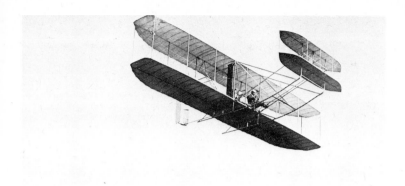

The Wright aeroplane in flight over Dayton, Ohio, c. 1904

Tail sections of the B-17 F bomber ready for assembly, Seattle, Washington, during World War II

Enrico Fermi, 1901–1954

Dr J. Robert Oppenheimer, 1904–1962

Side', with its reference to the new chemical weapons and world-wide destruction, articulated that sense of disillusionment for a new generation. And the nation that celebrated its achievements in song and used the song to sell its products, now used it as a weapon to castigate itself. It was, and is, a measure of the nation's cynicism. The Bomb, after all, was an American invention, though the premise—that a uranium nucleus split in half would release a huge amount of energy—was a European one.

America's acceptance of it was perhaps her last great act of scientific dependence on the brains of the Old World. In 1939, President Roosevelt requested America to explore this possibility of creating a new kind of energy. That same year, under the guidance of the Italian immigrant nuclear physicist Enrico Fermi, the first experiments were performed in America to demonstrate the reality of nuclear fission—that energy is produced when a neutron collides with the nucleus of a uranium atom.

By 1940, with the pooled knowledge of the Germans Hahn, Frisch and Meitner, the Dane, Neils Bohr, the Hungarian Szilard, the French Joliot-Curies, the British Cockcroft and the American scientists themselves, it was known that with each collision, other neutrons—between one and three—are produced to cause further collisions in a split-second chain reaction. This energy could, perhaps, be harnessed for peaceful purposes. Or released for war. The catalyst that caused the political and scientific chain reaction that finally produced the atomic bomb was Japan's attack on Pearl Harbor in December 1941. On 13 August 1942, the American government established the 'Manhattan Engineer District'—the Manhattan Project—to develop the Bomb.

Four months later, Enrico Fermi built the world's first atomic nuclear reactor—the 'pile'—under the football stands of the University of Chicago. 'Pile', because it consisted of piles of uranium, uranium oxide and blocks of graphite. And it worked. The chain reaction became self-sustaining. Scientific theory became scientific fact. A new age—the Atomic Age—an Italian discovery as significant for America and the world as the discovery four and a half centuries earlier by that other Italian, Columbus, of the continent of America itself.

In the spring of 1943, at Los Alamos in the deserts of New Mexico, America established a physics research laboratory to work on the construction of the Bomb. In charge, as Fermi was an alien and technically an enemy, J. Robert Oppenheimer. On 16 July 1945, both men witnessed the world's first atomic explosion. The research cost was well over two billion dollars and the US

War Department called it 'the outstanding achievement of nuclear science'.[1] It is ironical that the notion of the research team, envisaged by Edison to produce 'inventions' for the benefit of mankind, should have as its greatest achievement the production in record time of a weapon of destruction.

Hiroshima, 6 August 1945. Nagasaki, 9 August 1945. America had her revenge for Pearl Harbor. There followed, on an island in the Pacific in 1952, the hydrogen bomb—brainchild of Hungarian-American scientist Edward Teller. Oppenheimer fought for international control, but it was too late. He was dubbed by Senator Joe McCarthy 'a security risk'. Teller spoke against him. The Bomb escalated within a decade to a power equivalent to 2500 times that of the bomb on Hiroshima, and the world was faced with the terrifying prospect of annihilation.

In the light of our ultimate survival, we should stop work on nuclear research. In the light of our being able to maintain a standard of living that is now based on the exploitation of natural resources, we cannot afford to. Raw materials are running out. The struggle to find alternative sources of energy—nuclear energy—becomes more crucial. And the temptation to use nuclear energy to grab what world resources are left, becomes greater.

Science has not given man more self-control, more kindliness, or more power of discounting their passions in deciding upon a course of action

The hydrogen bomb exploded by the US Atomic Energy Commission in the Marshall Islands in the autumn of 1952. Elugelab Island was completely destroyed. The cloud stem pushed upwards 25 miles, deep into the stratosphere, and the mushroom spread for 100 miles. The photograph was taken at 12,000 feet, 50 miles from the detonation site.

H-Bomb protest, New York, April 1958

. . . that is why science threatens to cause the destruction of our civilisation.[146]

Even if we can restrict ourselves to research into the peaceful uses of nuclear energy, the attendant risks are enormous—greater than they have ever been. The automobile gave us car accidents, the nuclear power station accident gives us fall-out. And, already, in sealed containers buried in the sea is enough radioactive waste from the first Bomb to kill life on Earth at any time during the next million years. Maybe it's a risk we have to take—a risk made inevitable by the society and economy that's been created by the American System of Manufacture. Like the South in 1800, we seem locked into a system that is self-generating and which it will be difficult to escape from. The South escaped through the holocaust of the Civil War. Maybe war will be the only way out for us.

> I'm going to preach you a sermon 'bout Old Man Atom,
> And I don't mean the Adam in the Bible datum,
> I don't mean the Adam Mother's Eve mated,
> I mean the thing science liberated.
> You know, Einstein said he was scared,
> And if he's scared, I'm scared.[147]

Equal with the Bomb as the most powerful symbol of 'science out of control', is the computer. The machine takeover. The angel of disenchantment with our modern industrial society. We joke about its inefficiency because we feel threatened by it.

The development of the computer parallels the development of almost every other major American innovation, from its conception in Europe to its exploitation in the States. Charles Babbage (1792–1871), British mathematician, was founder of the Cambridge Analytical Society. Eighty years before Taylor and Ford, he worked on the concept of the division of labour in factories. He invented the skeleton key, the speedometer, and a British version of the cow-catcher, but he is most important not for what he achieved but for what he anticipated. As early as 1822, and in a society that was increasingly machine-orientated, he began to consider how machinery might be used 'for the purposes of computation'. Not simple adding up, but a system whereby one could mechanically check astronomical and other mathematical data, the errors of which 'littered the log tables and contemporary text books'. 'I wish to God,' he said, 'these calculations could be executed by steam.' It was the decade of George Stephenson's 'Rocket' and the opening of the Liverpool-Manchester railway.

Charles Babbage, 1792–1871

Babbage's scheme was financed by an unusually far-sighted British government who could see the value of a sophisticated mechanical calculator, particularly in wartime, for making nautical tables for the Royal Navy. In 1823, the Royal Society, chaired by Humphrey Davy, expressed its encouragement:

It appears that Mr Babbage has displayed great talents and ingenuity in the construction of his machine for computation.[148]

But, by 1842, George Airy, the Astronomer Royal advising the government, took this position:

I replied, entering fully into the matter, and giving my opinion that it was worthless.

All that the 'great talent' had produced was a half-realised leviathan—an 'Analytical Engine' which was to become Babbage's obsession. Babbage was handicapped in the practical realisation of his ideas by contemporary technology, by the lack of gears and levers sufficiently advanced to cope. But the concept itself was for a general-purpose computer almost in the modern sense. A machine that would use perforated cards to store both data and partial answers:

Thus the Analytical Engine will possess a library of its own. Every set of cards once made will at any future time reproduce the calculations for which it was first arranged. The numerical value of its constants may then be inserted.

Already-solved and stored answers could have additional work done on them and the answers be printed automatically. 'We may say most aptly,' wrote Lady Lovelace, Byron's daughter and Babbage's backer, 'that the Analytical Engine weaves algebraic patterns.'[149] But almost thirty years later, in 1870, Babbage could only say 'the machine is not constructed yet, but I am working at it'. In 1871, Babbage died. The grandfather of computers; a man before his time.

The American engineer Herman Hollerith (1860–1929) was probably unaware of Babbage's work, but where Babbage had failed, recent advances in technology enabled Hollerith to succeed. In place of cumbersome mechanical levers, Hollerith was able to use electricity, and in the late 1880s produced his Tabulating Machine.

Hollerith was activated by the desire, usual for Americans in the era of Edison, to make a fortune. He achieved it by devising a new method for counting and analysing the results of a census. Head counting in the old days, when America had a population of a few million, was comparatively simple. Now, with a population

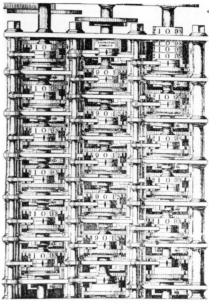

One of Babbage's calculating machines, the Difference Engine, started in 1823 and abandoned in 1842

Herman Hollerith, 1860–1929

approaching sixty million, not so simple. Counting the returns by hand was impossibly long.

My idea at that time was to use a strip of paper and punch the record for each individual in a line across the strip. Then I ran this strip over a drum and made contact through the holes to operate the counters.[16]

An electrical circuit was completed only if a hole was present. A complete circuit moved counters (dials) one place for each hole. With a series of counters, each representing one particular characteristic of the population—male or female, black or white—it was easy to extract the answers to specific, if fairly basic questions. A good idea, but unworkable. Rolls of punched paper would operate player-pianos, but to record facts about Americans Hollerith would need, in his own words, 'miles of paper'. He turned to punched cards—more manageable, more easily stacked, more quickly processed. Babbage had used 'operation cards', having taken the idea from the way the Frenchman Jacquard used punched cards to automate the weaving process on his looms. But Hollerith, according to his son, got the idea from the American railroads.

At that time in the West, so many train robbers posed as passengers that the Government asked the railroads to keep a record of everybody aboard. To do this, the railroads used tickets as identification cards. When the conductor first collected a ticket, he punched holes in it by those items that described the passenger in a printed list. This was called, I believe, a punch-photograph. . . . I often heard my father tell about watching the conductor use a hand punch to mark his ticket—light hair, dark eyes, large nose, medium height—and realizing suddenly that his tabulating cards should be punch-photographs of each person in the census.[16]

Hollerith had no hand-out from the Government. It was still the land of individual opportunity; he conducted his own personal financial struggle, backing his own brains and, in 1885, producing his first prototype, with cards the size of a dollar bill and space for 288 holes.

He tried out his machine with the help of the Baltimore City Fathers, who allowed him to process the city's death records. Into thousands of cards he punched holes according to the deceased's age, sex and cause of death, and proved that his machine could do in days what it took clerks weeks. In 1890 he competed for the government census contract. The competition took on overtones of Elias Howe and the seamstresses, as Hollerith, with his machine, raced against an army of tally clerks with pen and ink. Hollerith completed the work twice as quickly as his nearest rival and, unlike Howe, was awarded the contract. Thirty machines at $1000 each, and each machine with interchangeable parts and cards of identical proportions. A mere six weeks after the first census returns, Hollerith announced the American population as $62\frac{1}{2}$ million:

The Eleventh Census handled the records of 63,000,000 people and 150,000 minor civil divisions. One detail (characteristic) alone required the punching of one billion holes. Because the electrical tabulating system of Mr Hollerith permitted easy counting, certain questions were asked for the first time. Examples of these were:
Number of children born
Number of children living
Number of family speaking English
By use of the electric tabulating machine it became possible to aggregate from the schedules all the information which appeared in any way possible. Heretofore such aggregations had been limited. With the machines, complex aggregations can be evolved at no more expense than simple ones.[150]

It was a triumph of accuracy, of technology, of the American passion for speed. And Hollerith was rewarded with the traditional fruits of American success: the yacht, the motor-cars, the elegant house, the adoring wife and a place in the country.

In 1911, Hollerith sold his Tabulating Machine Company to an outfit that in 1924 changed its name to International Business Machine Company. Today's ubiquitous IBM. But the machine remained, basically, a calculator. It never wove the 'algebraic patterns' conceived by Babbage. It was the businessman's super-clerk. Pen pushers gradually becoming button pushers.

In the following year, 1925, an American electrical engineer, Vannevar Bush (born 1890), built a machine that was able to solve differential equations. The great Lord Kelvin, transatlantic cable layer, Centennial judge and supporter of Bell's telephone, had designed such a machine on paper fifty years before, but Bush's machine was the first analogue computer and the first since Babbage to store information with the notion of performing additional calculations on it later. During World War II, the progeny of these machines reached such a degree of sophistication that the American government was able to use them for making the complicated calculations associated with the Manhattan Project. It was the brilliant mathematical brain of John von Neumann (1903–57) that solved for the physicists the problem of how to create the fast reaction essential for the great explosion. But von Neumann's greatest contribution to the Manhattan Project was to show the scientists how to express their

Dr Vannevar Bush, 1890–

The Electronic Numeric Integrator and Calculator. One of its first calculations in nuclear physics would have taken one hundred engineers, using conventional methods, a year to solve. The ENIAC produced the answer in two hours.

theories mathematically so that the computer could solve the problems for them. The result was the world's most modern computer laboratory and the vindication of the British Government's initial faith in Babbage's machine as a convenience for war. Sadly for Britain, it came one hundred years late—13 August 1942—and on the wrong side of the Atlantic.

In 1946, at a dedication ceremony for the ENIAC computer—500 multiplications per second, 5000 additions—Major General Barnes, Head of Research and Development, saw the machine as:

. . . the means of extending the frontiers of knowledge with all that implies for the betterment of mankind.

But, in the small print, the rider:

. . . every effort will be made to permit the great potential usefulness of this great scientific tool to be realized as broadly as possible . . . within the limitations imposed by the requirements of national security.[150]

Even the computer—tool of peace, tool of war. And the free exchange of ideas that characterised much of science in the nineteenth century, gone, hidden behind the curtains of 'national security'.

In the same year, 1946, Bush produced the first electronic

digital computer.* Mechanical switches were replaced by electronic ones. It still used punched cards but had an electronic storage unit—a memory bank. In the 1950s, the power unit, the vacuum tube, was replaced by Shockley's transistor, and computers became smaller. Today, electronic impulses are being replaced by laser beams that record information on strips of plastic.

The computer enables scientists to make routine calculations that would otherwise take impossibly long. For example, the calculations that the German Renaissance astronomer Kepler needed to do in order to work out the orbit of the planet Mars took him four years. In 1964, a computer did the same calculations in eight seconds. Computers and their uses have proliferated. They can calculate the structural stress of a skyscraper whilst it's still on the drawing board, tabulate flight data, defend the nation against surprise attack, guide guided missiles, send men to the moon. Itself the product of mass-production, the computer can control the mass-production system: automate an entire factory. It has heralded a second Industrial Revolution, where men are no longer subordinate to the system but irrelevant to it.

I can conceive of an orchard operated entirely without workers in which the fruit-bearing trees can be cared for and cultivated and the fruit picked by machines operated and controlled by electronic devices. . . . The inventive mind can and probably will develop farming into an entirely new type of activity in which the physical burdens are lifted from the shoulders of the farm workers.[130]

Irving Langmuir, crystal gazing in the 1940s—and with the best of humanitarian intentions. But our lurking paranoia about a machine take-over increases. And with the astronomical cost of a computer such that it must be kept at work, the American System of Manufacture has arrived at a point where machines are more valuable, in economic terms, than men.

The computer industry's theoretician was Norbert Wiener (1894–1964). His wartime work had been a study of anti-aircraft defence: how to shoot down planes—ensure a direct hit—by calculating almost instantaneously the plane's speed, direction, height, the strength of the wind and the curve and speed of the land-based missile. Computers were the answer. Wiener summarised his conclusions in his book, *Cybernetics*, the term des-

*Definitions: 'Analogue' Computer (Americans read 'Analog')=calculating machine that operates by representing numerical magnitudes by such physical quantities as voltages.
'Digital' Computer=calculating machine that operates by representing numbers as digits in the decimal or some other system.

cribing 'the science of control and communication in the animal and the machine', the equation of the machine with the human nervous system. It became the computer Bible. In 1947, Wiener, like Oppenheimer and even Langmuir, refused to contribute further to military research and devoted himself to warning mankind of the dangers of automation:

It is perfectly clear that this will produce an unemployment situation in comparison with which . . . the depression of the thirties will seem a joke . . . thus the new industrial revolution is a two-edged sword. It may be used for the benefit of humanity, assuming that humanity survives long enough to enter a period in which such benefit is possible. If, however, we proceed along the clear and obvious lines of our traditional behavior, and follow our traditional worship of progress and the fifth freedom—the freedom to exploit—it is practically certain that we shall have to face a decade more of ruin and despair.[151]

Mass-production—conformity—a punched-card society—automated social system—form-filling bureaucracy—the dehumanisation of the individual. The London *Times*, a small item, 24 March 1975:

The growing use of computers to store information about a person's credit worthiness and other personal details has led to widespread anxiety about possible misuse. For example, should the information be wrong it could lead to injustice for the rest of a person's life if uncorrected.

Legal curbs on the use of data banks were promised soon. But there remains the threat of social misuse, the centralisation of control, the manipulation of the populace by governments. And the 'game' theory. Von Neumann, at the same time as he was helping the Manhattan Project, was writing *The Theory of Games and Economic Behaviour* (published in 1944). It showed the best tactics to be employed in simple games—and not so simple. The same tactics, said von Neumann, can be applied to those more serious games of business and war by working out mathematically the best way to outsmart a business rival, destroy an enemy.

> How beauteous mankind is! O brave new world
> That has such people in't.

And power to them that control the switches.

Such liberal ideas as free trade, free press, unbiased education, either already belong to the past or soon will do . . . it would seem probable that, in the next fifty years or so, we shall see a still further increase in the power of governments, and a tendency for governments to be such as are desired by the men who control armaments and raw materials.

The forms of democracy may survive in western countries, since those who possess military and economic power can control education and the press, and therefore can usually secure a subservient democracy.[146]

'O Brave New World.' New York 1973— high rise, dwarfed people, gin and tobacco. The cigarette ad blew smoke rings every 30 seconds.

Bertrand Russell's 'in the next fifty years' is our today.

Paradoxically, it is the computer that might be the very tool to help us avoid the dangers that already seem to be with us. 'Information is data that changes us.' The modern computer can select from its memory bank the data that is most important to the task at hand, calculate probabilties, test our hypotheses. Von Neumann argues that it is possible to construct a machine more intelligent than ourselves. Hopefully, he's right, as the computer could be the only instrument we have, able to cope with the incredible complexity of the 'multitudinous interdependent and interacting changes' that man has to struggle with. The computer cannot predict our future but it can, fed the right information and correctly programmed, indicate the way that future might go— offer us a choice of action.

In early America, the needs of the nation were simple and obvious. Today, they are neither obvious nor simple, and they must be seen in a world context. The American System of Manufacture itself became a product for export. And the bosses went with it. Europe colonised the world geographically—

Urban sprawl—Interstate Highway 10, Santa Monica Freeway, California

Precision Engineering—electronic telephone switching system. Magnetic 'bubbles' (large white dots) are moved through a shift register. The bubbles are 4 thousandths of an inch in diameter.

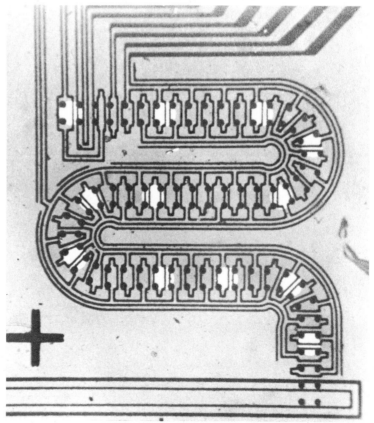

America economically. But the US is no longer immune to the
upheavals in the rest of the world. The self-sufficiency she strove
for and won she is losing again—witness oil.

Reduced to as simple terms as possible and cyberneticised,
America is faced with what most people already recognise but are
reluctant to accept: a world in which, by 1990, the population
and industrial pollution will double, a fall in food output will lead
to mass starvation, natural resources will be depleted, the danger
of nuclear war will increase, totalitarianism will replace democracy
and Western society collapse. What the French President M.
Giscard d'Estaing termed 'the world's revenge on Europe [and
America] for the nineteenth century'. Against this . . . by controlling the birth rate, controlling pollution, giving priority to
food production, sparingly using non-renewable resources and
recycling wherever we can, the situation can be stabilised and
Armageddon deferred. Even avoided for ever.

The Age of Invention, as understood in the first one hundred
and fifty years of America's history, is over. We may not like
what has been invented. But we can't un-invent it. And it happened so quickly that only now are we beginning to realise what
has been created: the American System of Manufacture is an
enormity, like a Hollywood King Kong. And we must now make
the necessary mental as well as physical adjustments that will
enable us to deal with it.

America's challenge for the future is to use technology to solve
the problems of technology without producing a further technological escalation. It could mean a slowing down of mass-
production, an end to planned obsolescence, an end to the
expected flow of new goodies—which don't always turn out to be
so good, *vide* thalidomide; and a start to long-range economic
planning.

Some warn that planning will lead to socialism but I think it's worth
the risk.

And if Henry Ford junior thinks the free-enterprise system is not
working properly, then Ford is probably right.

In my thirty years as a businessman, I have never felt so uncertain
about the future of my country and my company.

It will mean a more equitable share-out of the world's food and
the world's natural resources. President Ford, Detroit, 1974:

No one can foresee the extent of the damage nor the end of the disas-
trous consequences if nations refuse to share nature's gifts for the
benefit of all mankind.

Which means not just Arab oil to America, but an end to America's, and to a lesser extent Europe's colossal consumption of everything produced by everybody else. It will mean an end to the automatic assumption that our standard of living should increase; and a start to building man back into the system that threatens to destroy him. The problem is to reconcile the known remedy with each American individual's individualism and his urge to compete and be a success.

There is, significantly, no technology-worshipping exhibition to celebrate the American Bicentennial. But, if there were one and there were a President Grant to open it, his 1976 speech would reflect no easy optimism. He would acknowledge the real achievements of the past and outline the challenge for the future —a challenge facing the whole of Western technological society, of which Europe was the cradle and the United States of America the fulfilment.

A challenge much greater than the opening up of the West. A Global Challenge. A challenge to use technology to maintain our standard of living whilst using our native inventiveness to preserve some of those freedoms that technology has forced us to give up. Freedoms which brought America into being. Freedoms which helped make her the greatest industrial nation on Earth. Freedoms which are enshrined in her mythology but which, in real terms, have ceased to exist. New freedoms: which are really the old freedoms in different guise—freedom from war, freedom from the pollution of industry, freedom from the tyranny of the machine, freedom from want and hunger. Not only in America—but in the whole World.

We have built our nation and our nation is no longer young. Our past experience has shown that we can solve our problems. Man may not be infinitely wise—but he has proved himself infinitely resourceful.

Sign at Walt Disney World, Florida, 1975

Roll of Honour

For those Early American inventors
who should have made this book
but didn't

1750 Jacob Yoder, Penn. Flatboat for inland rivers
1787 John Fitch, Penn. Steamboat, *The Thornton*
1787 James Rumsey, Md. Steamboat
1790 Jacob Perkins, Mass. Nail cutter
1799 Eliakim Spooner Seeding machine
1809 Abel Stowel, Mass. Screw cutting machine
1812 Peter Gaillard, Penn. Grass cutter
1816 George Clymer, Penn. Hand printing press, 'The Columbian'
1819 Ezra Dagget Canned oysters
1821 Sophia Wood Straw hat
1822 C. M. Graham Artificial teeth
1825 Thomas Blanchard, Mass. Steam automobile
1826 William Kendall Insertable teeth for rotary saws
1830 Joseph Dixon, Mass. Lead pencil
1832 Egbert Egberts, N.Y. Power knitting machine
1832 John Matthews, N.Y. Soda-water apparatus
1835 Henry Brandon, Vt. Horseshoe machine
1836 Alonzo D. Phillips, Mass. Friction match
1838 Alfred Vail, N.J. Printing telegraph
1839 Charles Goodyear, Conn. Vulcanised rubber
1840 Jonas Chickering Grand piano
1841 Edwin Chaffee, Mass. India rubber ball
1842 Crawford W. Long, Ga. Ether as anaesthetic
1843 Enos Wilder Fireproof safe
1844 J. and S. Battin, Penn. Anthracite coal breaker
1844 A. D. Puffer, Mass. Soda-water cooler
1848 Henry P. Westcott Wooden peg machine
1849 Henry Evans, N.J. Pendulum press for can tops
1850 Dr John Gorrie, Fla. Compressed air ice machine
1852 Christopher Dorflinger, N.Y. Lamp chimneys
1852 Channing and Farmer, Mass. Electric fire alarm
1856 Gail Borden, Conn. Condensed milk
1859 Henry E. Steinway, N.Y. Improved grand piano
1862 E. A. L. Roberts, W.Va. Torpedo for oil drilling
1865 William Bullock, Penn. Web printing press
1869 John Hyatt, N.Y. Plastic for billiard balls
1874 William Baldwin The 'Bundy' radiator
1875 Volney Barkley Corn kerneling machine
1883 Scott and Chisholm Automatic pea sheller
1886 Charles Hall Commercial aluminium
1888 Frank Sprague, Va. Overhead-wire electric trolley system

and one modern inventor
1928 Vladimir Zworykin Iconoscope, TV camera

Book List

There are a considerable number of books which the reader may find useful. The two books which I have used throughout are:

1 Wilson, M. A., *American Science and Invention*, New York: Simon and Schuster, 1954.
2 Oliver, J. W., *History of American Technology*, New York: Ronald Press, 1956.

I have also referred to volumes of *Scientific American*, *Harper's Magazine*, *Chamber's Journal*, the London *Times*, and *Atlantic Monthly*. All other books are listed below under the chapter in which they are first used. The footnote numbers in the text refer to the numbers in this list. Some quotations appear in more than one source, in which case I have referred to the original source.

Chapter 1

3 Boorstin, D. J., *The Americans: the National Experience*, New York: Random House, 1965; London: Weidenfeld and Nicolson, 1966.
4 Struik, D. J., *Yankee Science in the Making*, New York: Macmillan, 1962.
5 Franklin, B., *The Works of Dr Benjamin Franklin*, Edinburgh: Oliver and Boyd, 1812.
6 Franklin, B., *The Papers of Benjamin Franklin*, vols 1–18, Yale University Press, 1964–74.
7 Franklin, B., *Autobiography*, ed. L. Labaree *et al.*, Yale University Press, 1964.
8 Van Doren, C. C., *Benjamin Franklin*, Toronto: Macmillan, 1938; Westport, Conn.: Greenwood, n.i. 1973.
9 Seeger, R. J., *Benjamin Franklin: New World Physicist*, Oxford: Pergamon Press, 1973.
10 Priestley, J., *The History and Present State of Electricity*, London, 1767.
11 Scudder, H. E., ed., *Recollections of Samuel Breck*, Philadelphia: Porter and Coates, 1877; Kelley, USA, n.i.
12 'Correspondence of Eli Whitney relevant to the Invention of the Cotton Gin', *American Historical Review*, 1897.
13 Green, C. M., *Eli Whitney and the Birth of American Technology*, Boston: Little, Brown, 1956.
14 Mirsky, J. and Nevins, A., *The World of Eli Whitney*, New York: Collier Books, 1962.
15 Woodbury, R. S., 'The Legend of Eli Whitney and interchangeable parts', *Technology and Culture*, New York, 1960.
16 *Those Inventive Americans*, Washington: National Geographic Society, 1971.
17 Bruchey, S. W., ed., *Cotton and the Growth of the American Economy, 1790–1860: sources and readings*, New York: Harcourt, Brace and World, 1967.
18 White, G. S., *Memoir of Samuel Slater*, Philadelphia, 1836; Kelley, USA, n.i., 1970.
19 Appleton, N., *The Introduction of the Power Loom, and Origin of Lowell*, Lowell, Mass.: B. H. Penhallow, 1858.
20 Miles, H. A., 'The Growth of Lowell' (1846), reprinted in Glaab, C., *The American City*, Homewood, Ill.: Dorsey Press, 1963.

21 Miles, H. A., *Lowell, as it was and as it is*, 1845; New York: Arno, n.i. 1972.
22 Jefferson, T., *The Writings of Thomas Jefferson*, ed. P. L. Ford, 10 vols, New York: Putnam, 1893–9.
23 Stanford, C. L., 'The Intellectual Origins and New Worldliness of American Industry', *Journal of Economic History*, 1958.
24 Dickens, C. *American Notes*, London: Chapman and Hall, 2 vols, 1842; Oxford University Press, 1957; Penguin Books, 1972; New York: Dutton, 1970.
25 Larcom, L., *A New England Girlhood*, Boston: Houghton Mifflin, 1889; New York: Corinth Books, facsim. ed., 1970.
26 Weld, C., *A Vacation Tour in the United States and Canada*, London, 1855.
27 *Genius Rewarded: the Story of the Sewing Machine*, New York: Singer Sewing Machine Company, 1880.
28 James Parton, article in *Atlantic Monthly*, 1867.
29 *New York Industrial Exhibition. Special reports of Sir Joseph Whitworth and George Wallis*, London, 1854.
30 Burlingame, R., *March of the Iron Men: Social History of Union*, New York: Scribner, 1939.

Chapter 2

31 Boorstin, D. J., *The Americans: the Democratic Experience*, New York: Random House, 1973.
32 Brown, D., *Bury my Heart at Wounded Knee*, New York: Holt, Rinehart and Winston, 1971: Bantam, 1972; London: Barrie and Jenkins, 1971; Pan, 1973.
33 Holloway, D., *Lewis and Clark and the Crossing of North America*, London: Weidenfeld and Nicolson, 1974.
34 Merk, F. and L. B., *Manifest Destiny and Mission in American History*, New York: Knopf, 1963; Vintage Books, 1973
35 Dillon, J. G. W., *The Kentucky Rifle*, Washington, D.C.: National Rifle Association, 1924.
36 Edwards, W. B., *The Story of Colt's Revolver*, Harrisburg, Pa.: Stackpole Co., 1953.
37 Russell, C. P., *Guns on the Early Frontiers*, University of California Press, 1957.
38 Prescott Webb, W., *The Great Plains*, New York: Grosset and Dunlap, 1931; paperback, 1957.
39 Thompson, C. N., *Sons of the Wilderness: John and William Conner*, Indianapolis: Indiana Historical Society, 1937.
40 Casson, N. H., *C. H. McCormick, his Life and Work*, Chicago: McClurg, 1909.
41 Hutchinson, W. T., *Cyrus Hall McCormick*, New York: Century Co., 1935; Da Capo, 2nd ed., 2 vols, 1969.
42 Rodgers, C. T., *American Superiority at the World's Fair*, Philadelphia, 1852.
43 Bathe, G. and D., *Oliver Evans: a Chronicle of Early American Engineering*, Historical Society of Pennsylvania, 1935; Arno, n.i. 1972.
44 Marryat, F., *A Diary in America*, London: Longman, 3 vols, 1839; Indiana University Press, n.i. 1960; Westport, Conn.: Greenwood, n.i. 1973.

45 Weld, C., '*A Vacation Tour in the US and Canada*', 1855.

46 Thoreau, H., '*Walden*', 1845.

47 Morse, E. L., ed. *Samuel F. B. Morse: His Letters and Journals*, Boston: Houghton Mifflin, 1914.

48 Mabee, F. C., *The American Leonardo: a Life of Samuel F. B. Morse*, New York: Knopf, 1944; Octagon, n.i. 1969.

49 Prime, S. I. *Life of Samuel F. B. Morse*, New York: Appleton, 1875; Arno, n.i. 1974.

50 Coulson, T., *Joseph Henry, his Life and Work*, Princeton University Press, 1950.

51 Dickenson, E. N., *Joseph Henry and the Magnetic Telegraph*, 1885.

52 *Revolution in Transport*

53 Holbrook, S. H., *Story of American Railroads*, New York: Crown, 1947.

54 Bowles, S., '*Our New West*' (1869), reprinted in Hawgood, J. A., *The American West*, London: Eyre and Spottiswoode, 1967.

55 Lavender, D., *The Penguin Book of the American West*, London: Penguin Books, 1969.

56 Macaulay, J., *Across the Ferry: First Impressions of America and its People*, London, 1871.

57 Husband, J., *The Story of the Pullman Car*, Chicago: McClurg, 1917; Arno, n.i. 1972.

58 Mencken, A., *The Railroad Passenger Car*, Johns Hopkins U.P., 1957; O.U.P., 1958.

59 Rae, W. F., *Westward by Rail*, London: Isbister, 1870; New York: Arno, n.i. 1973.

60 Leupp, F. E., *George Westinghouse: His Life and Achievements*, Boston: Little, Brown, 1918; London: John Murray, 1919.

61 Prout, H. G., *A Life of George Westinghouse*, New York: Scribner, 1922; Arno, n.i. 1972; London: Benn, 1922.

62 McCoy, J. G., *Historic Sketches of the Cattle Trade of the West and Southwest*, Glendale, Cal.: A. H. Clark, 1874; University Microfilms, n.i. 1966.

63 Kipling R., *From Sea to Sea: Letters of Travel*, London: Macmillan, 2 vols, 1900.

Chapter 3

64 Helper, H. R., *The Impending Crisis of the South: How to meet it*, New York, 1857; Harvard University Press, n.i. 1969.

65 Bruce, R. V., *Lincoln and the Tools of War*, Indianapolis: Bobbs-Merrill, 1956; Westport, Conn.: Greenwood, n.i. 1974.

66 The diary of John Hay, private secretary to President Lincoln, quoted in Bruce, *op. cit.*

67 Strong, G. T., *Diary of the Civil War, 1860–65* ed. A. Nevins, New York: Macmillan, 1962.

68 Rogers, H. C. B., *The Confederates and Federals at War*, London: Ian Allan, 1973.

69 *Harper's Magazine*, June 1886.

70 Sleeman, C. W., *Torpedoes and Torpedo Warfare*, Portsmouth: Griffin, 1880.

71 Dickinson, H. W., '*Robert Fulton: Engineer and Artist*', London, 1913, reprinted in introduction to Sleeman, *op. cit.*

72 Fulton, R., *Torpedo War and Submarine Explosions*, New York: W. Eliot, 1810; Chicago: Swallow Press, n.i. 1971.

73 Robinson, W. M., *The Confederate Privateers*, Yale University Press, 1928.

74 Barnes, R. H., *United States Submarines*, New Haven, Conn.: H. F. Morse, 1944.

75 Haydon, F. S., *Aeronautics in the Union and Confederate Armies*, vol. 1, Johns Hopkins University Press, 1941; Oxford University Press, 1941.

76 De Vries, L., *Victorian Inventions*, London: John Murray, 1971.

Chapter 4

77 Maass, J., *The Glorious Enterprise: the Centennial Exhibition of 1876 in Philadelphia*, New York: American Life Foundation, 1973.

78 United States Centennial Commission International Exhibition, *Official Catalogue*, New York, 1876.

79 *New York Times*, 15 May 1876.

80 Richards, G. T., *The History and Development of Typewriters*, London: H.M.S.O., 1964.

81 Harrigan, E., *The Telephone*, 1876, unpublished (New York City Library and Mrs Joshua Logan).

82 Bell, A. G., *Visible Speech as a means of communicating articulation to deaf-mutes*, Washington: Gibson Bros., 1872.

83 Bruce, R. V., *Bell: Alexander Graham Bell and the Conquest of Solitude*, London: Gollancz, 1973.

84 Mackenzie, C. D., *Alexander Graham Bell*, Boston: Houghton Mifflin, 1928.

85 Dickson, W. K. L. and A., *The Life and Inventions of Thomas Alva Edison*, London: Chatto and Windus, 1894.

86 Josephson, M., *Edison: a Biography*, New York: McGraw-Hill, 1959; paperback, 1963.

87 Edisonia Ltd, *A Catalogue of Phonographs*, 1898; City of London Phonograph, n.i. 1966.

88 *Chamber's Journal*, August 1888.

89 Wile, F. W., *Emile Berliner: Maker of the Microphone*, Indianapolis: Bobbs-Merrill, 1926; New York: Arno, n.i. 1974.

90 Gelatt, R., *The Fabulous Phonograph*, New York: Lippincott, 1955; London: Cassell, 1956.

91 Burlingame, W. R., *Out of Silence into Sound*, New York: Macmillan, 1964.

92 Muybridge, E., *Animal Locomotion*, Philadelphia: University of Pennsylvania, 1887; Da Capo, n.i. no date.

93 Ackerman, C. W., *George Eastman*, Boston: Houghton Mifflin, 1930; Kelley, USA, n.i. 1935; London: Constable, 1930.

94 Coe, B., *George Eastman and the Early Photographers*, London: Priory Press, 1973.

Chapter 5

95 Wayne-Morgan, H., *Unity and Culture, the United States 1877–1900* (Pelican History of the USA, Vol 14), London: Penguin Books, 1974.

96 Williamson, H. F. and Daum, A. R., *The American Petroleum Industry*, New York: Northwestern University Press, 2 vols, 1959 and 1963.

97 Clark, J. S., *The Oil Century: from the Drake Well to the Conservation Era*, University of Oklahoma Press, 1958.

98 Leander Bishop, J., *A History of American Manufactures from 1608 to 1860*, vol. 2 *Petroleum*, Philadelphia: E. Young, 1868; Kelley, USA, n.i. 1966.

99 Leonard, C. C., *History of Pithole*, 1867.

100 Giddens, P. H., *The Birth of the Oil Industry*, New York: Macmillan, 1938; Arno, n.i. 1972.

101 Giddens, P. H., *Early Days of Oil*, Princeton U.P.; 1948; Peter Smith, n.i. 1965.

102 Dolson, H., *The Great Oildorado*, New York: Random House, 1959.

103 *Erie Weekly Dispatch*, 28 January 1868, quoted in Giddens, *Early Days*.

104 Boatright, M. C., *Folklore of the Oil Industry*, Dallas, Texas: Methodist University Press, 1963.

105 Boatright, M. C., *Tales from the Derrick Floor*, New York: Doubleday, 1970.

106 Tarbell, I. M., *The History of the Standard Oil Company*, New York: McClure, 1904; abbr. ed. Norton, 1969.

107 Myers, G., *History of the Great American Fortunes*, 3 vols, Chicago: Charles Kerr, 1910.

108 Abels, J., *The Rockefeller Millions*, London: Muller, 1967.

109 Nevins, A., *John D. Rockefeller: the Heroic Age of American Enterprise*, New York: Scribner, 1940; Kraus Reprints, 2 vols, no date.

110 Greenleaf, W., *Monopoly on Wheels: Henry Ford and the Seldon Automobile Patent*, Detroit: Wayne State University Press, 1961.

111 Taylor, F. W., *The Principles of Scientific Management*, New York: Harper, 1911; Westport, Conn.: Greenwood, n.i. 1972; Norton, n.i. 1967.

112 Ford, H. and Crowther, S., *My Life and Work*, New York: Doubleday, 1922; Arno, n.i. 1973: London: Heinemann, 1922.

113 Sward, K., *The Legend of Henry Ford*, New York: Rinehart, 1948; Russell, n.i. 1968.

114 Nevins, A., *Ford: the Times, the Man, the Company*, New York: Scribner, 1954.

115 Burlingame, R., *Henry Ford: a Great Life in Brief*, New York: Knopf, 1955; Quadrangle, 1970; London: Hutchinson, 1957.

116 *Horseless Age*, vol. 6, no. 20.

117 *Scientific American*, 1899.

118 Glaab, C., *The American City*, Homewood, Ill.: Dorsey Press, 1963.

119 'Burnaby's Travels through North America', New York, 1904; quoted in Glaab, *op. cit.*

120 Chevalier, M., 'Society, Manners and Politics in the United States', 1835; quoted in Glaab, *op. cit.*

121 Tucker, G., 'Progress of the United States in Population and Wealth', New York, 1843; quoted in Glaab, *op. cit.*

122 Kingsbury, F. J., 'The Tendency of Men to live in Cities', *Journal of Social Sciences*, 33; quoted in Glaab, *op. cit.*

123 Kouwenhoven, J., *Columbia Historical Portrait of New York*, New York: Doubleday, 1953.

Chapter 6

124 *Strand Magazine*, 1897.

125 Gunston, D., *Guglielmo Marconi: Father of Radio*, London: Weidenfeld and Nicolson, 1965.

126 Letter in possession of the Marconi company.

127 De Forest, L., *Father of Radio*, Chicago: Wilcox and Follett, 1950.

128 Barnouw, E., *A History of Broadcasting in the United States*, vol 1. *A tower in Babel*, New York: Oxford University Press, 1966.

129 Dunlap, O. E., *Radio's 100 Men of Science*, New York: Haper, 1944.

130 *Langmuir, the Man and the Scientist*, Parts 1 and 2, by A. Rosenfeld and J. H. Hildebrand, Oxford: Pergamon Press, 1962.

131 Kettering, C., *Biographical Memoir of Leo Hedrik Baekland*, Washington: National Academy of Science, 1942.

132 Address by Leo Baekland in *The Journal of Industrial and Chemical Engineering*, Vol. 1, New York, 1909.

133 Gunther, J., *Inside USA*, London: Hamish Hamilton, 1947.

134 Dutton, W. S., *Du Pont: One hundred and forty years*, New York: Scribner, 1942.

135 Beynon, H., *Working for Ford*, London: Alan Love, 1973.

136 Sorensen, C. E., *Forty years with Ford*, London: Cape, 1957.

137 Leonard, J. N., *Tragedy of Henry Ford*, New York: G. P. Putnam & Sons, 1932.

138 Richard Wilbur, 'Death of a Toad', Faber and Faber, London.

Chapter 7

139 Gibbs-Smith, C. H., *The Aeroplane*, London: HMSO, 1960.

140 Kelly, F. C., *The Wright Brothers*, London: George Harrap, 1944.

141 Kelly, F. C., ed., *Miracle at Kitty Hawk: the Letters of Wilbur and Orville Wright*, New York: Straus and Young, 1951; Arno, n.i. 1971.

142 *Flying* magazine, 1924.

143 Chanute, O., 'Progress in Flying Machines', *American Engineer and Railroad Journal*, New York, 1894.

144 Langley, S. P., *Langley Memoir on Mechanical Flight*, Part 1 *1887–96*, Washington: Smithsonian Institute, 1911.

145 Asimov, I., *Biographical Encyclopedia of Science and Technology*, New York: Doubleday, 1964.

Picture Credits

146 Russell, B., 'The Effect of Science on Social Institutions', *Survey Graphic*, April 1924.

147 Old Man Atom by Verne Partlow and Irving Bibo. By permission of Cromwell Music Ltd, c/o Essex Music Ltd, 19/20 Poland Street, London, England.

148 Moseley, M., *Irascible Genius: a Life of Charles Babbage, Inventor*, London: Hutchinson, 1964.

149 Moulton, Lord, *The Invention of Logarithms. Its genius and growth*, London, 1915.

150 Goldstine, H. H., *The Computer, from Pascal to von Neumann*, Princeton University Press, 1972. Reprinted by permission of Princetown University Press.

151 Wiener, N., *The Human Use of Human Beings: Cybernetics and Society*, Boston: Houghton Mifflin, 1950; Avon, 1967; London: Eyre and Spottiswoode, 1951; Sphere, n.e. 1969.

Acknowledgement is due to the following for permission to reproduce pictures.

Titlepage, Solomon D. Butcher Collection, Nebraska State Historical Society; page 4, George Eastman House; 7 (top), U.S. Patent Office; 7 (bottom), Culver Pictures Inc.; 8, Maryland Historical Society; 12, Bettmann Archive; 13, Radio Times Hulton Picture Library; 14, Franklin, B., *Experiments and Observations on Electricity*; 15, Radio Times Hulton Picture Library; 17, Franklin, B., *Experiments and Observations on Electricity*; 18, Radio Times Hulton Picture Library; 19, Bettmann Archive; 20, Brown Bros; 21 (top), Radio Times Hulton Picture Library; 21 (bottom), Library of Congress; 22, Mabel Brady Garvan Collection, Yale University Art Gallery; 25, Bettmann Archive; 26, Bettmann Archive; 27, Astor, Lenox and Tilden Foundations, Prints Division, New York Public Library; 28, Bettmann Archive; 31, Astor, Lenox and Tilden Foundations, Prints Division, New York Public Library; 32, Bettmann Archive; 33, Bettmann Archive; 34 (top), Bettmann Archive; 34 (bottom), Library of Congress; 35, Singer Co. (UK) Ltd; 36, Radio Times Hulton Picture Library; 37, Library of the American Antiquarian Society; 38 (left), Bettmann Archive; 38 (right), Brown Bros; 39 Photo-Illustrators, Philadelphia; 40, Montana Historical Society, Helena; 42, Radio Times Hulton Picture Library; 44 (top), Bettmann Archive; 44 (bottom), Radio Times Hulton Picture Library; 45, Bettmann Archive; 46, Library of Congress; 47, Bettmann Archive; 48, Conner Prairie, Pioneer Settlement; 49, Deere & Company; 50, Deere & Company; 51, International Harvester; 52, International Harvester; 53, International Harvester; 54, Deere & Company; 55, Library of Congress; 57, Culver Pictures Inc.; 58, Staten Island Historical Society; 59 (top), New York Daily Tribune, 1849; 59 (bottom), Minnesota Historical Society; 60 (top), Bequest of Moses Tanenbaum, 1937, Metropolitan Museum of Art; 60 (bottom left), Bettmann Archive; 60 (bottom right), Union Pacific Railroad Museum; 61 (right), Union Pacific Railroad Museum; 62, Bettmann Archive; 63, Radio Times Hulton Picture Library; 65 (top), Union Pacific Railroad Museum; 65 (bottom), Burndy Library; 66 (top), Culver Pictures Inc.; 66 (bottom), Bettmann Archive; 67, Denver Public Library Western Collection; 68 and 69, Union Pacific Railroad Museum; 71 (top), Southern Pacific Railroad; 71 (centre), Bettmann Archive; 71 (bottom), Culver Pictures Inc.; 72 (top), Denver Public Library, Western History Department; 73, New York Historical Society; 74, Bettmann Archive; 76, S. D. Butcher Collection, Nebraska State Historical Society; 77, Bettmann Archive; 78, Kansas State Historical Society; 79, New York Historical Society; 81, U.S. Patent Office; 82, New York Historical Society; 83, New York Historical Society; 84, Brady Collection, Library of Congress; 87 (top), Leslie, F., *Pictorial History of the War*; 87 (bottom), Museum of the Confederacy; 88, Brady Collection, Library of Congress; 89, Smithsonian Institution; 91 (top), Culver Pictures Inc.; 91 (bottom left), Bettmann Archive; 91 (bottom right), U.S. Navy Photograph, Library of Congress; 92, Bettmann Archive; 93, Radio Times Hulton

Index